SPACE DEBRIS

SPACE DEBRIS
Hazard Evaluation and Mitigation

Edited by

Nickolay N. Smirnov
Moscow M.V. Lomonosov State University, Moscow, Russia

CRC Press
Taylor & Francis Group
Boca Raton London New York

CRC Press is an imprint of the
Taylor & Francis Group, an **informa** business
A TAYLOR & FRANCIS BOOK

CRC Press
Taylor & Francis Group
6000 Broken Sound Parkway NW, Suite 300
Boca Raton, FL 33487-2742

First issued in paperback 2019

ISSN 1026-2660
ISBN-13: 978-0-415-27907-9 (hbk)
ISBN-13: 978-0-415-27907-9 (pbk)

Visit the Taylor & Francis Web site at
http://www.taylorandfrancis.com

and the CRC Press Web site at
http://www.crcpress.com

Typeset in Singapore by Scientifik Graphics (Singapore) Pte Ltd

Every effort has been made to ensure that the advice and information in this book is true and accurate at the time of going to press. However, neither the publisher nor the authors can accept any legal responsibility or liability for any errors or omissions that may be made. In the case of drug administration, any medical procedure or the use of technical equipment mentioned within this book, you are strongly advised to consult the manufacturer's guidelines.

British Library Cataloguing in Publication Data
A catalogue record for this book is available from the British Library.

Library of Congress Cataloging in Publication Data
A catalog record has been requested.

Cover illustration: Artist's conception of debris cloud surrounding the Earth. (Courtesy Centre Pilot, 2000)

CONTENTS

PREFACE

Fundamental ecology is a branch of science aimed at developing mathematical models that could forecast the impact of technogeneous processes on the natural environment. The present edition illustrates the methods and models of fundamental ecology taking the outer space contamination problem as an example.

Since the first Sputnik was launched on 4 October 1957 and the space era began mankind was enthusiastic about putting satellites into orbit, using the wonderful opportunities given by the space achievements for telecommunications, navigation, Earth observations, weather forecasts, microgravity science and technology, etc., and nobody gave thought to a possible negative impact on the space environment. Now it is high time we step aside and look around.

Space activity of mankind generated a great deal of orbital debris, i.e. manmade objects and their fragments launched into space, inactive nowadays and not serving any useful purpose. Those objects, ranging from microns up to decimeters in size, traveling at orbital velocities, remaining in orbits for many years and numbering billions formed a new media named 'space debris' and became a serious hazard to space flights. Collision with a metallic particle of debris with 1 cm radius is energetically equivalent to a collision with a car moving at a speed of 100 km per hour.

Thus this media wherein the space satellites operate nowadays should be taken into account, and its impact on the durability of space missions should be evaluated as it will affect the reliability of the technical systems. That turns out to be of tremendous importance for developing the systems containing constellations of low Earth orbiting satellites as a space segment. The prospective 'SkyBridge' system, a satellite-based broadband access system which will provide high speed Internet access, videoconferencing and other services is supposed to be based on 64 low Earth orbit satellites. Developing the concepts of such systems it is necessary to take into consideration the space debris environment the systems will operate in. (Now the chosen altitude for the Skybridge constellation is 1457 km which practically coincides with the second maximum of the orbital debris population.)

Space debris has been studied by the leading scientific and research institutions of all the space powers since the '80s, and its implications have been discussed among the narrow circles of specialists at the international conferences and meetings of the space agencies from the '90s. But now the emerging constellations of the low Earth orbiting communication satellites or systems providing interactive broadband services and the forthcoming construction of the International Space Station brought the attention of a wide spectrum of specialists to the space debris problem.

The aims of the present edition are:

- to introduce the problem of space debris to a wide circle of people who are consumers of the space operating systems;
- to give a description of the existing level of space contamination and forecasts for future growth of the orbital debris population;
- to provide brief coverage of the activities on space debris issues that have been undertaken and are continuing;
- to emphasize the aspects affecting the growth of the space debris population in low Earth orbits and geosynchronous orbits;
- to present the modern methods of orbital debris evolution modeling, analysis and collision risks assessments;
- to describe the major hazards for the operating space systems and possible preventive measures;
- to discuss the mitigation strategies.

The contributors to the book are well-known specialists in the field of orbital debris studies from Europe, Japan, Russia and the USA, who have been in this business for many years. Most of them have experience of both working in research corporations or space agencies and being professors in universities. They were among the first to join the space debris community, their pioneer publications on the issue are well known.

The book is addressed to a wide spectrum of readers. An unfamiliar reader will find new concepts introducing the problems of space ecology, new data and examples. A specialist will find actual data on space debris environment, new mathematical models for space debris evolution, production and self-production, description of the existing software, concepts for shield design. A person involved in developing concepts for multifunctional constellations of low Earth orbit satellites will find methods of collision risks assessments depending on altitudes and inclinations of orbits, collision hazards evaluations and suggestions concerning preventive measures. The book will be suitable as a primary (or supplementary) text for college courses and in-service training programs.

CONTRIBUTORS

P.D. Anz-Meador
Lunar and Planetary Institute
Houston, Texas, USA

V.A. Chobotov
The Aerospace Corporation
El Segundo, California, USA

W. Flury
ESA/ESOC
Darmstadt, Germany

A.B. Kiselev
Faculty of Mechanics and Mathematics
Moscow M.V. Lomonosov State University
Moscow, Russia

A.I. Nazarenko
Center for Space Observations
Russian Space Agency
Moscow, Russia

V.F. Nikitin
Faculty of Mechanics and Mathematics
Moscow M.V. Lomonosov State University
Moscow, Russia

A.E. Potter
Lunar and Planetary Institute
Houston, Texas, USA

N.N. Smirnov
Faculty of Mechanics and Mathematics
Moscow M.V. Lomonosov State University
Moscow, Russia

T. Yasaka
Department of Aeronautics and Astronautics
Kyushu University
Hakozaki, Fukuoka, Japan

CHAPTER 1

ORBITAL DEBRIS HAZARDS ASSESSMENT AND MITIGATION STRATEGIES

V.A. Chobotov

The present Chapter gives a description of Space debris environment in Low Earth Orbits and in the Geosynchronous Orbit. Collision hazards, orbital breakups and further debris clouds evolution are discussed. The possible mitigation strategies are analyzed. A brief coverage of Interagency/ International activities on the Space debris issues is given.

1.1. Space Debris Environment

1.1.1. Introduction

Past design practices and inadvertent explosions in space have created a debris population in operationally important orbits. The debris consists of spent spacecraft and rocket stages, separation devices, and products of explosions.

Much of this debris is resident at altitudes of considerable operational interest. Two types of space debris are of concern: 1) large objects whose population, while small in absolute terms is large relative to the population of similar masses in the natural flux (by a factor of about 1000); and 2) a large number of smaller objects whose size distribution approximates natural meteoroids. Products larger than 10 cm^2 in low orbits can be observed directly. The existence of a substantially larger population of small fragments can be inferred from terrestrial tests in which the particle distributions from explosions have been assayed. Statistical measurements and returned surfaces from space confirm the existence of a large number of very small particles in orbit.

Man-made space debris differs from natural meteoroids because it is in Earth orbit during its lifetime and is not transient through the regions of interest. As a consequence, a given mass of material presents a greater challenge in the design and operation of spacecrafts because of the extended time period over which there is a risk of collision. Future satellite designs should take into account the space debris in addition to the natural environment consistent with mission requirements and cost effectiveness.

Some efforts to provide a better assessment of the orbiting debris problem have been and are being made by various government agencies and international organizations. Principal areas of investigation are the hazards related to the tracked (cataloged), untracked, and future debris populations. Studies are being conducted in the areas of technology, space vehicle design, and operational procedures. Among these are ground- and space-based detection techniques, comprehensive models of Earth-space environment, spacecraft designs to limit accidental explosions, and different collision-hazard assessment methods. Occasional collision avoidance and orbit-transfer maneuvers are being implemented for selected satellites. The results and experience gained from the activities will, in time, create a better understanding of the problem and all its implications so that appropriate actions can be taken to maintain a relatively low-risk environment for future satellite systems.

Impact protection from space debris may not be feasible in most cases because of very high approach velocities and the fact that certain protuberances, especially those of relatively large areas such as solar arrays and antennas, cannot easily be shielded permanently. Evasive-maneuvering techniques may reduce the present probability of collision for specific satellites in certain circumstances.

The only natural mechanism opposing debris buildup is removal by atmospheric drag. This process can take a very long time, however, especially from high altitudes, and causes debris to migrate from higher to lower altitudes. Another mechanism, collection by a spacecraft, would be extremely difficult and very expensive. Prevention of debris formation is the most effective approach.

At the present time, the collision hazard is real but very low relative to other hazards of space operations. This hazard can be minimized by initiating studies and implementing their results in five major areas: 1) education, 2) technology, 3) satellite and vehicle design, 4) operational procedures and practices, and 5) national and international space policies and treaties (*ed.* by Chobotov, 1996).

1.1.2. Space Debris Environment: Low Earth Orbit (LEO)[*]

At any one time, there are about 200 kg of meteoroid mass moving through altitudes below 2000 km at an average speed of about 20 km/s. Most of the mass is found in particles of about 0.1-mm diameter. The meteoroid environment has always been a design consid-

---

[*] *Orbital Mechanics*, 2nd edition, edited by V.A. Chobotov – Copyright © 1996 AIAA – Reprinted with permission.

eration for spacecraft. The Apollo and Skylab spacecraft were built to withstand impacts on critical systems from meteoroids having sizes up to 3 mm in diameter. Larger sizes were so few in number as to be of no practical significance for the duration of the mission. Some small spacecraft systems required additional shielding against meteoroids as small as 0.3 mm in diameter in order to maintain an acceptable reliability. The trend in the design of some future spacecraft (as for example, the International Space Station) is toward larger structures, lighter construction, and longer times in orbit. These factors increase the concern about damage from particles in the 0.1- to 10-mm size range.

It is no longer sufficient, however, to consider only the natural meteoroid environment in spacecraft design. Since the time of the Apollo and Skylab programs, launch activity has continued and increased. As a result, the population of orbital debris has also increased substantially. The total mass of debris in orbit is now approximately 2300 tons at altitudes below 2000 km. Relative to one another, pieces of debris are moving at an average speed of 10 km/s, or only half the relative speed of meteoroids. The significant difference between the orbital debris population and the meteoroid population is that most of the debris mass is found in objects several meters in diameter rather than 0.1 mm in diameter as for meteoroids. This large reservoir of mass may be thought of as a potential source for particles in the 0.1- to 10-mm range. That is, if only one ten-thousandth of this mass were in this size range, the amount of debris would exceed the natural meteoroid environment. The potential sources for particles in this size range are many:

1) Explosions: More than 145 spacecraft are known to have exploded in LEO and account for about 42% of the U.S. Space Command (USSPACECOM) Catalog.
2) Hypervelocity collision in space: One known satellite collision (CERISE) and several near misses have been recorded in recent history.
3) Deterioration of spacecraft surfaces: Oxygen erosion, ultraviolet radiation, and thermal stress are known to cause certain types of surfaces to deteriorate, producing small particles. Returned surfaces from orbit (SOLAR MAX, LDEF, SFU, EuReCa, Euro Mir '95 and others) have provided some information on this effect.
4) Solid rocket motor firings: Up to one-third of the exhaust products of a solid rocket motor may be aluminum oxide particles in the size range 0.0001 to 0.01 mm. Slag (up to several centimeters in diameter) may be generated during and after the burn of solid rocket motors.
5) Unknown sources: Other sources are likely to exist. Particulates are commonly observed originating from the Space Shuttle and other objects in space.

a) Debris Measurements

What is currently known about the orbital debris flux is from a combination of ground-based and in-space measurements. These measurements have revealed an increasing population with decreasing size. Beginning with the largest sizes, a summary of these measurements follows.

The USSPACECOM tracks and maintains a catalog of "all man-made objects" in space. The catalog, as of 30 September 1998, contained 8668 objects, most in LEO.

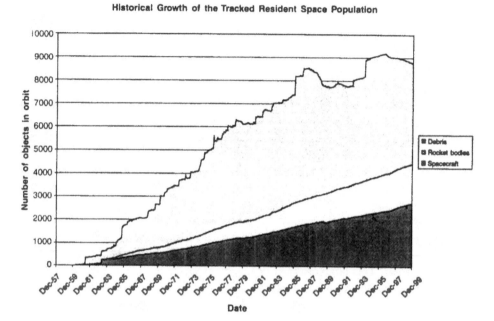

Fig. 1.1. Historical growth of the tracked resident space population (Stacked chart).

Fig. 1.1 shows the growth of the satellite population from 1957 through 1999. This plot excludes space probes (over 100 of which were still in orbit). The linear growth rate of about 210 cataloged objects in the last 40 years provides a good approximation of the actual growth rate which is a function of satellite breakup rates and cyclic solar activity.

The origin of the satellite and debris population is given in Table 1.1 and its composition is illustrated in Fig. 1.2. As can be seen from Fig. 1.2, nearly half of the objects in the catalog have resulted from fragmentation of more than 145 satellite breakups. The

Table 1.1. Orbital Origins (31 December 1998) (Source: NASA, The Orbital Debris Quarterly News, January 1999).

(as of 31 December 1998, as catalogued by US SPACE COMMAND)			
Country/Organization	Payloads	Rocket, Bodies & Debris	Total
China	24	101	125
Russia	1340	2579	3919
ESA	24	213	237
India	17	4	21
Japan	65	49	114
US	828	3139	3967
Other	266	25	291
Total	2564	6110	8674

Approximate Catalog Composition

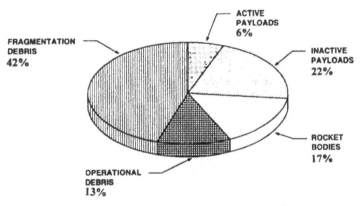

Fig. 1.2. Approximate catalog composition.[*]

ability to catalog small objects is limited by the power and wavelength of individual radar sites, as well as the limitations on data transmission within the network of radar sites. Consequently, objects smaller than 10 to 20 cm are not usually cataloged.

An estimate of USSPACECOM's capability to detect objects in Earth orbit is illustrated in Fig. 1.3. Only the region to the right of the heavy line is accessible to operational radar

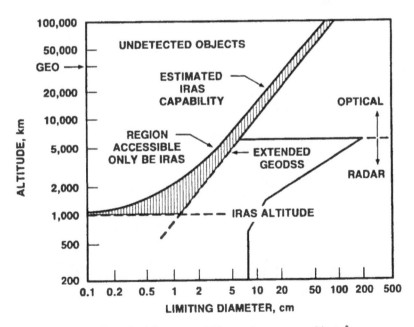

Fig. 1.3. United States capability to detect space objects.[*]

[*] *Orbital Mechanics*, 2nd edition, edited by V.A. Chobotov – Copyright © 1996 AIAA – Reprinted with permission.

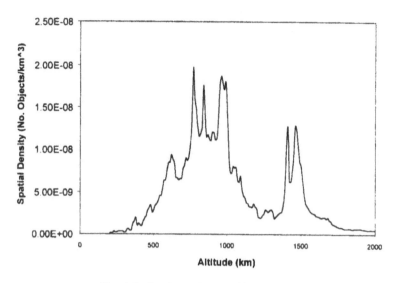

Fig. 1.4. Catalog object density in LEO.

and optical systems. The capabilities of the infrared astronomy satellite (IRAS) extended the measurements to smaller objects, as indicated.

The tracked (>10-cm) object densities are illustrated in Fig. 1.4. The peak object densities appear at about 800-, 1000-, and 1500-km altitude and are caused by heavy use of these altitudes. A number of high-intensity explosions or breakups of spacecraft and rocket stages have also contributed to the debris population in this environment. The inclination distribution of the catalog population is shown in Fig. 1.5.

Nearly all of the orbital debris measurements to date show an orbital debris flux that exceeds the meteoroid flux. Cross-sectional flux of a given size and larger is given and compared with the meteoroid flux in Fig. 1.6.

b) Future Debris Population Estimates

Currently there are a number of models, which predict future orbital debris environments such as the NASA's EVOLVE, the ESA's CHAINEE and MASTER, the UK's IDES and the Russian Space Agency models. Given certain assumptions these and other models predict an exponential rise in the orbital debris population. Principal uncertainties involve future launch rates and object intercollision rates which would produce more fragments than would decay from atmospheric drag. A key question in this regard is the estimation of the number of fragments from a collision or an explosion. Improved breakup models are necessary to reduce the uncertainties in this area. Consequently, the future space debris environment will depend on:

1. on-orbit breakups (explosions/collisions)
2. non-fragmentation debris sources
3. constellation deployments
4. models of future traffic

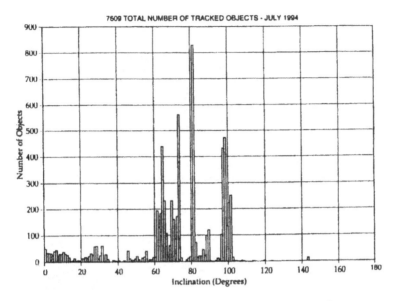

Fig. 1.5. Inclination distribution of catalog population.*

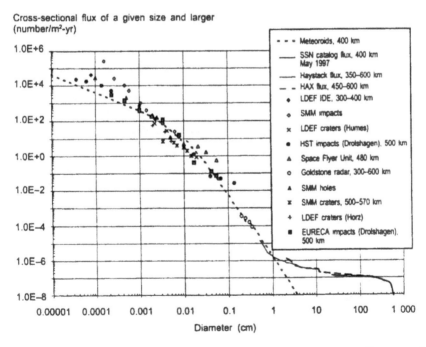

Fig. 1.6. Orbital debris measurements compared to meteoroid flux (Technical Report on Space Debris – United Nations, New York, 1999).

---

* *Orbital Mechanics*, 2nd edition, edited by V.A. Chobotov – Copyright © 1996 AIAA – Reprinted with permission.

Table 1.2. Geosynchronous items in USSPACECOM Catalog

Owner	Spacecraft	Rocket bodies	Debris
United States	1200	20	3
Soviet Union/CIS	132	129	2
Great Britain	7	—	—
Italy	3	—	—
Canada	11	—	—
France	9	—	—
Australia	5	—	—
Japan	37	1	—
Germany	5	—	—
NATO	8	—	—
PRC	19	—	—
India	12	—	—
Spain	3	—	—
European Space Agency	19	1	1
France/Germany	3	—	—
Indonesia	10	—	—
ITSO	54	—	—
Brazil	6	—	—
Saudi Arabia	7	—	—
Mexico	5	—	—
Sweden	4	—	—
Luxembourg	10	—	—
International Maritime	9	—	—
Thailand	3	—	—
Total	581	151	6

1.1.3. Space Debris Environment: Geosynchronous Orbit (GEO)

Orbits in the geosynchronous regime may be classified as 1) low or high inclination, and 2) sub- or supersynchronous-altitude drifting orbits. The ideal geostationary orbit has an altitude of 35, 787 km and a period of 1436.2 min. A satellite whose period of revolution is equal to the period of rotation of the Earth about its axis is called a *geosynchronous satellite*. A geostationary satellite is a geosynchronous satellite in a circular orbit that lies in the plane of the Earth's equator and rotates about the Earth's polar axis in the same direction. Consequently, it remains "fixed" over a point on the equator.

As of the end of 1996 there were a total of 509 payloads launched into the geosynchronous orbit, not including 186 upper stages. Of these there were some 250 spacecraft (functional and non-functional) in the geostationary orbit. The others are either geosynchronous or highly elliptical (Molnyia) semisynchronous orbits.

A representative distribution of spacecraft and rocket bodies in GEO is shown in Table 1.2. The population density peaks at the geostationary altitude and decreases sharply

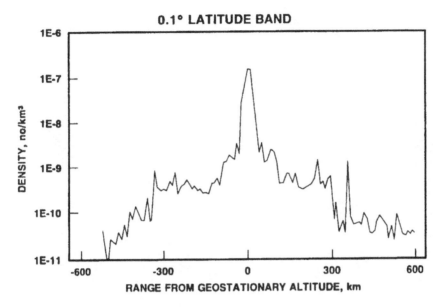

Fig. 1.7. Population density as a function of range and from GEO.

with distance from the geostationary altitude. This is illustrated in Fig. 1.7 with the effect of latitude also shown.

GEO orbital debris is more difficult to detect and track than debris at lower altitudes. Consequently, GEO debris characterization lacks detail in small objects and fragmentation and breakup debris. Plans for an international effort in GEO debris characterization are being prepared involving US, ESA, Japanese, Russian and Chinese facilities to conduct GEO orbital debris surveys. A debris population in GEO down to 10 cm from the present 1 m limiting size would provide a more accurate object density and enhance the estimates of the collision probabilities for active and large inactive satellites in GEO.

1.2. Collision Hazards

1.2.1. Collision Probability

The hazards can be broadly defined in two categories; hazards due to explosion and hazards due to collision. Explosions are a concern because of the probabilities of collision of other satellites with the debris from the explosion. For example, on 3 June 1996 the STEP II Mission Pegasus/HAPS upper stage (Hydrazine Auxiliary Propulsion System) exploded in an orbit of 586 km by 821 km with an inclination of 82.0°. The mass of the HAPS was estimated as 97 kg which produced approximately 700 debris objects identified and tracked by the U.S. Space Surveillance Network (SSN). The debris spread from 250 km to more than 2500 km.

Collisions are also a concern because an unplanned collision may damage an active spacecraft and because of the collision probabilities of other satellites with the debris.

The collision of the French CERISE spacecraft and a fragment from the SPOT 1 Ariane I upper stage on 24 July 1996 produced only one new piece of debris, a portion of a boom. This event marked the first time that two objects in the U.S. satellite catalog have collided (*The Orbital Debris Quarterly News*, September 1996). This collision with CERISE's stabilization boom caused the spacecraft to tumble end-over-end. Collisions with smaller particles that are below the threshold for detection and tracking are more frequent and also can cause serious damage.

The assessment of the probability that any two objects will collide in orbit can be made by a number of methods as follows:

a) Poisson Distribution

Based on the kinetic theory of gases, the Poisson distribution model yields a simple method for computing the probability of collision between a satellite and any one of the other objects in orbit. This approach is accurate if the population of objects can be assumed to resemble random motion of molecules in a gas which is induced by repeated collisions.

The Poisson distribution approach has been used by the NASA engineering models for orbital debris and meteoroids. It has also been used by the NASA environmental prediction model, EVOLVE, the ESA MASTER model, the Debris Environment and Effects Program (DEEP), and the Russian Space Agency debris contamination model (Kessler *et al.*, 1989; Kessler, 1981).

By the Poisson distribution method the probability of collision between two objects is given by the relation

$$P(col) = 1 - e^{-N_{enc}} \qquad (1.1)$$

$$N_{enc} = \int_0^{t_f} \rho v_r A_c dt \qquad (1.2)$$

A satellite with a projected area A_c, moving with a mean relative velocity v_r will sweep out a volume $V = v_r A_c \Delta t$ in a time increment Δt. The flux of objects is ρv_r, where ρ is the object density in V. For $\rho v_r \ll 1$, the probability of collision is aproximately given by

$$P(col) \approx \rho v_r A_c \Delta t \qquad (1.3)$$

For large values of ρv_r, equation (1.3) provides the expected number of objects to be encountered in time interval Δt.

*b) Distance of Closest Approach**

The probability that any two objects will collide is a function of the orbital parameters, object size, and time. However, the collision cannot take place unless the orbits approach

---

* *Orbital Mechanics*, 2nd edition, edited by V.A. Chobotov – Copyright © 1996 AIAA – Reprinted with permission.

each other within an effective collision radius R_s. This may occur even for initially nonintersecting orbits because the Earth's oblateness, air drag, and solar-lunar perturbations tend to alter the orbital parameters in time.

If it is assumed that the position uncertainties associated with the three dimensions (coordinates) of the nominal calculated miss distance between objects R_{min} are Gaussian (normal), with zero biases and equal variance, and are uncorrelated, a bivariate normal density function $f(x,y)$ can be defined in plane xy containing R_{min} that is oriented normal to the relative velocity vector at encounter. Thus,

$$f(x,y) = (2\pi\sigma^2)^{-1}\exp[-x^2/(2\sigma^2)]\exp[-y^2/(2\sigma^2)] \tag{1.4}$$

The probability of collision for $R_s \ll R_{min}$ takes the following form

$$P(col) = \iint f(x,y)dxdy = (2/\pi)(R_s/\sigma)^2 \exp[-R_{min}^2/2\sigma^2)] \tag{1.5}$$

c) Weibull Distribution

A probabilistic assessment of the collision risk can be determined directly by the use of Monte Carlo analysis in conjunction with a Weibull distribution fit to the random sample of the variable of interest. Random samples of a set of variables such as distances of closest approach are generated by numerical means, either by propagating trajectories of the objects of concern or by a Monte Carlo analysis in conjunction with methods of extreme order statistics. An analytical probability density function, i.e, a Weibull function, can be fitted to these data and theorems in extreme order statistics used to justify inference about probability of collision at the small extreme values of the minimum approach distances (Vedder and Tabor, 1991; Chobotov and Johnson, 1994; Chobotov *et al.*, 1997; Chobotov and Mains, 1998). The explicit formulation in this approach is as follows:

A Weibull probability distribution function fit to the absolute (global) minimum distances of closest approach is approximated in the form

$$f(x) = \left(\frac{\tau}{\beta}\right)\left(\frac{x}{\beta}\right)^{\tau-1} e^{-\left(\frac{x}{\beta}\right)^{\tau}} \tag{1.6}$$

Here, x is the random variable (i.e. minimum range); τ, β are the shape and scale parameters, respectively.

The cumulative Weibull collision probability function is of the form

$$F(x) = \int f(x)dx = 1 - \exp\left[-\left(\frac{x}{\beta}\right)^{\tau}\right] \approx \left(\frac{x}{\beta}\right)^{\tau} \text{ for } \frac{x}{\beta} \ll 1 \tag{1.7}$$

For example, consider the normalized relative minimum distance statistics between a given satellite and a set of resident space objects as shown in Fig. 1.8. Equation (1.6) (Weibull) fit is shown as a solid line. The cumulative probability distribution (equation

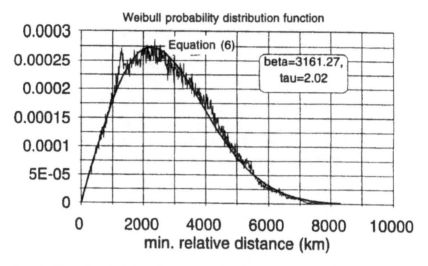

Fig. 1.8. Normalized relative distance statistics (60-second propagation over 6 month).

(1.7)) is given in Fig. 1.9. All close encounters are seen to occur within 8000 km and the peak encounter rate occurs near 2500 km.

The encounter rate N for an object at distance r_c, can be expressed mathematically as:

$$N = \frac{F(r_c)}{E(t_c)} \tag{1.8}$$

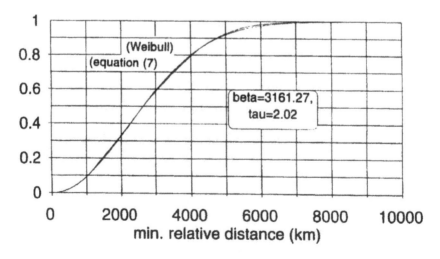

Fig. 1.9. Cumulative probability (60-second propagation over 6 month).

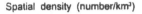

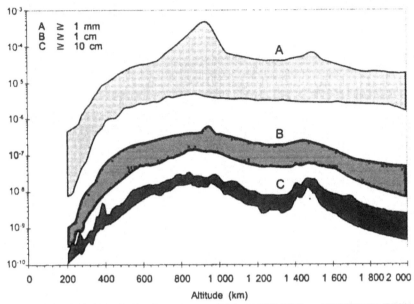

Sources: NASA (ORDEM96); DERA (IDES); ESA (MASTER); CNUCE (SDM); and NAZARENKO (SDPA)

Fig. 1.10. Model values for current spatial density (Technical Report on Space Debris – United Nations, New York, 1999).

where

$$F(r_c) = \int_0^{r_c} f(x)dx \qquad (1.9)$$

and

$$E(t_c) = \frac{E(c)}{v_r} \qquad (1.10)$$

Here $E(t_c)$ is the characteristic dwell time by an object within the radius r_c of the given spacecraft. $E(c)$ is the expected (mean) value of the chord of a circle with radius r_c and v_r is the mean relative velocity of objects. It can be shown that $E(c)$ is numerically equal to one fourth of the circumference of the circle of radius r_c.

The mean (expected) time to collision is then (Chobotov and Mains, 1998)

$$T_c = 1/N = (\pi r_c/2v_r)(\beta/r_c)^\tau \qquad (1.11)$$

An example of the Poisson distribution approach is illustrated in Fig. 1.10, where model

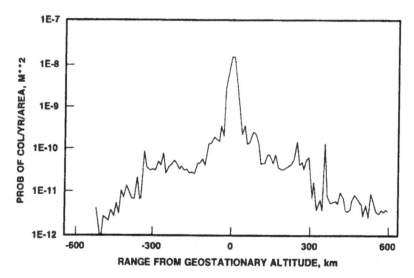

Fig. 1.11. GEO probability of collision.

values of spatial density for different particle sizes are illustrated for LEO. For GEO the collision probability per meter squared per year is shown in Fig. 1.11.

The probability of collision for a GEO satellite with cataloged debris using the distance of closest approach method is shown in Fig. 1.12.

Selected examples of the Weibull approach to compute the probability of collision can be found in (Vedder and Tabor, 1991; Chobotov and Johnson, 1994; Chobotov *et al.*, 1997; Chobotov and Mains, 1998).

1.2.2. Breakup Modeling

The fragmentation of upper stages and satellites accounts for nearly one-half of all trackable debris objects. Understanding the fragmentation process (due to explosion or hypervelocity collision) is therefore essential in order to estimate the mass, number, velocity, and ballistic coefficients of the resultant fragments. The modeling of the fragmentation debris (as orbiting clouds) can be separated into short- and long-term types. Short-term may be defined in terms of days after the breakup, while long-term implies subsequent evolution of the fragment clouds and the resulting "steady-state" environment due to the breakup. Primary elements of such modeling include: a) determination of breakup causes, b) orbital lifetime of untrackable debris, c) the breakup process (i.e., breakup modeling), d) debris cloud evolution, e) future traffic projections, and f) collision hazards to resident space objects. Fig. 1.13 illustrates the interfaces required between various types of models.

There are a number of empirically based fragmentation models that describe the number, mass, and velocity distributions of fragments resulting from hypervelocity collisions or explosions (see, for example, Jenkin, 1995; Chobotov, 1990; Chobotov *et al.*;

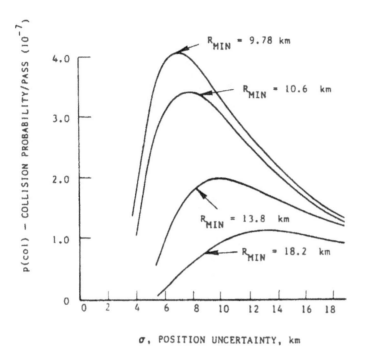

Fig. 1.12. Typical GEO collision probabilities as a function of position uncertainty and computed miss distance.*

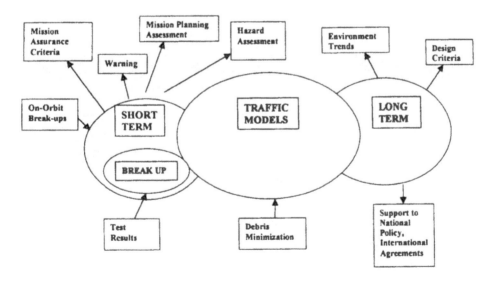

Fig. 1.13. Models.*

Chobotov and Spencer, 1991; Jenkin, 1993; Su and Kessler, 1984; McKnight, 1991; Sorge and Johnson, 1993; Wiedeman *et al.*, 1998). There is however, much uncertainty due to the limited experimental data that must be used to validate such models. Only more testing, coupled with the development of new analytical computational tools, can reduce these uncertainties. Even if those collisions are extremely rare it is important to develop a better understanding of the breakup phenomenology for the collisions between LEO objects. Such collisions are likely to occur with impact velocities in the 8- to 15-km/s range and result in hundreds of trackable fragments and potentially millions of smaller particles. The development of breakup models is essential. These models should include strict energy and momentum conservation laws such as are being implemented in program IMPACT, for example (Sorge and Johnson, 1993). Other models have been developed by numerical/ analytical approaches coupling existing models that describe local area effects due to hypervelocity impact (hydrocodes) with models that describe the spacecraft vibration and deformation due to the impact (structural response codes) (Spencer *et al.*, 1997).

a) Debris Cloud Evolution Modeling

The behavior of debris particles following a breakup in orbit must be modeled in order to determine the collision hazard for resident space objects in the vicinity of the breakup. Short-term behavior (measured in days) of a debris cloud for a hypervelocity collision is, for example, described in (Chobotov, 1990; Chobotov *et al.*; Chobotov and Spencer, 1991). Initially, the debris cloud dynamics model was developed based on the linearized rendezvous equations for relative motion in orbit. Subsequently, the short-term behavior

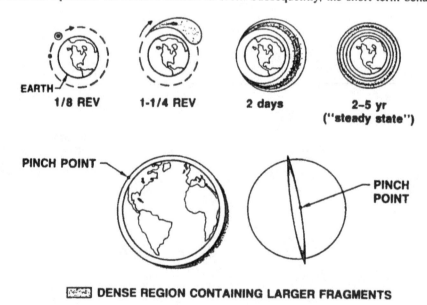

Fig. 1.14. Evolution of debris cloud.[*]

---

[*] *Orbital Mechanics*, 2nd edition, edited by V.A. Chobotov – Copyright © 1996 AIAA – Reprinted with permission.

and modeling of debris clouds in eccentric orbits were developed, which described the shape, in -track particle density variation, and volume of a debris cloud resulting form an isotropic breakup (Jenkin, 1993). A typical cloud density profile is shown in Fig. 1.14.

The DEBRIS model for example uses fragmentation data from IMPACT to simulate orbital debris cloud motion and determine the short-term collision hazard posed to a satellite (or a satellite constellation) operating near a recently formed debris cloud. During this time, regions of high spatial density, called constrictions or "pinch zones," are formed. For a given space vehicle, DEBRIS can compute the probability of collision and the associated mass, velocity, kinetic energy, momentum, and directionality distributions. The IMPACT and DEBRIS models were used to analyze the hazard from a Titan II upper-stage explosion. IMPACT modeled the explosion, and DEBRIS quantified the hazard to the Space Shuttle (STS-60) and the MIR space station which were operating with crews (Spencer *et al.*, 1997). Examples of the DEBRIS model output are shown in Figs. 1.15 and 1.16. These figures show the model's estimate for collision flux and probability encountered by the Space Shuttle and MIR Space Station as they operated near the expanding debris cloud.

1.3. Mitigation Strategies

About 13 percent of the present cataloged orbital debris population consists of objects discarded during normal satellite deployment and operations. The rest is the result of orbital breakups (42%), rocket bodies (17%) and inactive payloads (22%). Current miti-

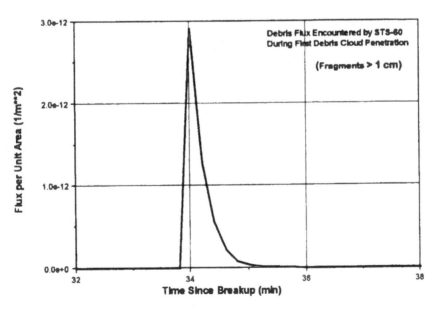

Fig. 1.15. Flux vs. time (DEBRIS output) (Spencer *et al.*, 1997).

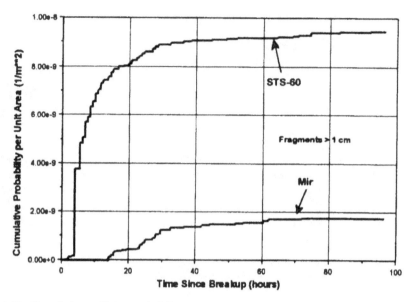

Fig. 1.16. Cumulative collision probability vs. time (DEBRIS output) (Spencer *et al.*, 1997).

gation strategies are therefore focused on limiting these sources of orbital debris in the future. They include:

1. Passivation of spacecraft and upper stages
2. Deorbiting of rocket bodies
3. Reorbiting of satellites to storage orbits
4. Collision avoidance and shielding technology

   The passivation methods include the expulsion of residual propellants by burning or venting the discharge of electrical storage devices, the release of pressurized fluids, thermal control and safing of unused destruct devices and the unloading of momentum wheels and similar attitude control devices.

   Deorbiting of rocket bodies and spent satellites is recommended in LEO at an end of mission or not to exceed a 25 year lifetime in orbit. For space objects at higher altitudes, moving (reorbiting) vehicles into disposal orbits can also be effective in the short term. For example, the transfer of geostationary orbit spacecraft at end of mission at least 300 km above GEO not only protects operational spacecrafts but also reduces the probability of derelict objects colliding with one another and creating debris that might threaten the GEO regime.

   The collision avoidance during orbital insertion and on-orbit operations is technically possible. However, current space surveillance systems cannot track objects in LEO with a radar cross section of less than 10 cm in diameter. In addition, it is difficult to maintain orbital parameters on small objects because of the higher area-to-mass ratio, and consequently, a higher dependence on the atmospheric density variations.

The United States Space Surveillance Network (SSN) and the Russian Space Surveillance System monitor the LEO environment to warn of an object approaching the Space Shuttle or the MIR space station within a few kilometers. Thus, if an object is predicted to pass through a box measuring 25 km × 5 km × 5 km oriented along the flight path of the Space Shuttle, the SSN sensor network intensifies its tracking of the potential collision object. If the improved fly-by prediction results in a conjunction within a box measuring 5 km × 2 km × 2 km, an avoidance maneuver may be performed. The Space Shuttle has performed four such evasive maneuvers since 1986.

A collision avoidance maneuver affects satellite operations in several ways and should be minimized consistent with spacecraft safety and mission objectives. The impact on propellant consumption, payload data and service interruptions and temporary reduction in tracking and orbit determination accuracy can be reduced with the improvement of tracking observation accuracy.

1.4. Inter-Agency/International Activities

1.4.1. Inter-Agency Activities

Within only a few decades, space has become an essential resource for science, defense and commercial utilization. These, as well as human activities in space, are however, increasingly at risk resulting from production of man-made space debris. The concern about this problem which causes a growing threat for the future of space flights has been recognized in the United States since the early 1980s. In 1987 the DoD Space Policy stated that "*...DoD will seek to minimize the creation of space debris in its military operations. Design and operations of space tests, experiments and systems will strive to minimize or reduce debris consistent with mission requirements.*" As a result of this policy an Interagency Group IG (Space) conducted a study in 1989 which recommended:

1. Make debris minimization a design consideration
2. Establish a DoD/NASA joint study group to develop plans for debris measurement, modeling and mitigation
3. Establish a DoD, NASA, DOT and commercial operation study to construct a research plan for developing technologies and procedures to mitigate debris

The U.S. National Space Policy on Orbital Debris was issued in 1989 which stated that "All space sectors will seek to minimize the creation of space debris. Design and operations of space tests, experiments, and systems will strive to minimize or reduce accumulation of space debris consistent with mission requirements and cost effectiveness. The U.S. government will encourage other spacefaring nations to adopt policies and practices aimed to debris minimization."

To carry out the IG (Space) report recommendations, a DoD/NASA study was conducted. That study provided more detailed goals for the research program and recommended that it consist of two phases. In the first phase, emphasis would be placed on increasing knowledge of the debris environment in LEO. The second phase would focus on determining the debris hazard and on improving spacecraft survivability. A program plan was submitted to the National Space Council and approved in July 1990. The DoD and NASA phase one program was completed in December 1993 (Spencer *et al.*, 1997). At this time DoD and NASA results were reviewed and an updated report was prepared under the auspices of the White House Office of Science and Technology Policy (OSTP).

The 1989 study was re-examined by the National Research Council (NRC) in 1993 at the request of NASA. A report issued by the NRC in 1995 recommended that:

1. Models of the future debris environment should be further improved. Uncataloged debris in LEO should be carefully studied.
2. Further studies should be conducted to better understand the GEO debris environment.
3. A strategy should be developed to gain an understanding of the sources and evolution of the small debris population.
4. The data acquired from this research should be compiled into a standard population characterization reference model.

In 1995, the OSTP released a report on space debris, "Interagency Report on Orbital Debris" (an update of a 1989 Inter-agency report). This report gave several recommendations on areas of space debris research and analysis:

• Continue and enhance debris measurement, modeling and monitoring capabilities;
• Conduct a focused study on debris and emerging LEO systems;
• Develop government/industry design guidelines on orbital debris;
• Develop a strategy for international discussions;
• Review and update U.S. policy on debris.

On September 14, 1996, President Clinton signed the latest National Space Policy. The policy states that *"The United States will seek to minimize the creation of space debris. NASA, the Intelligence Community, and the DoD in cooperation with the private sector, will develop design guidelines for future government procurements of spacecraft, launch vehicles, and services. The design and operation of space tests, experiments and systems, will minimize or reduce accumulation of space debris consistent with mission requirements and cost effectiveness. It is in the interest of the U.S. Government to ensure that space debris minimization practices are applied by other spacefaring nations and international organizations. The U.S. Government will take a leadership role in international fora to adopt policies and practices aimed at debris minimization and will cooperate internationally in the exchange of information on debris research and the identification of debris mitigation options."* The policy is consistent with the goals of the (International) Interagency Space Debris Coordination Committee (IADC), the United Nations' Committee on the Peaceful Uses of Outer Space (COPUOS), and other inter-governmental working groups.

A U.S. Government Orbital Debris Workshop for Industry was held in Houston on January 27-29, 1998. This workshop included 5 government agencies and representatives from the aerospace industry. Attendees expressed a desire for orbital debris mitigation standards and cited the need for a "level playing field" both domestically and internationally. The workshop suggested that standards need to be quantitative, explicit, and applied and enforced uniformly. Specific standards (guidelines) recommendations dealt with:

- solid rocket motor debris
- eliminating failure modes leading to explosions
- assuring postmission disposal
- collision avoidance for tethered systems
- favoring reentry over disposal orbits

An American Institute of Aeronautics and Astronautics Special Project Report on MEO/LEO Constellations was published (AIAA, 1999). The report focuses on the emerging legal regime for orbital debris mitigation.

1.4.2. International Activities

a) The International Academy of Astronautics (IAA)

The international community represented by the International Academy of Astronautics (IAA) published a position paper on space debris in 1992. Proposed methods of control included options that fall into three categories: those requiring minimal impact on operations, those requiring changes in hardware or operations and those requiring technology development. Options in the first category recommended for immediate application are given as follows:

1. No deliberate breakups of spacecraft which produce debris in long lived orbits.
2. Minimization of mission-related debris.
3. Safing procedures for all rocket bodies and spacecraft which remain in orbit after completion of their mission.
4. Selection of transfer orbit parameters to ensure the rapid decay of transfer stages.
5. Reorbiting of geosynchronous equatorial satellites at end-of-life (minimum altitude increase 300–400 km).
6. Upper stage and separated apogee kick motors used for geostationary satellites should be inserted into a disposal orbit at least 300 km above the geostationary orbit.

Second category options aim for removing used upper stages and dead spacecraft from orbit. This could be accomplished with deorbiting maneuvers to ensure atmospheric entry over ocean areas. Debris control options of category three require new developments where, in general, technical feasibility and cost effectiveness must be demonstrated.

Installation of drag enhancement devices and the use of lasers debris sweeps fall into this category.

The IAA Position Paper on space debris was published to update the factual information and re-examine some debris control options (IAA, 1995).

b) The Inter-Agency Space Debris Coordination Committee (IADC)

A preeminent international forum for exchanging technical information on orbital debris research and for coordinating joint studies has been established and is now comprised of ten members: China (CNSA), ESA, France (CNES), Germany (DLR), Great Britain (BNSC), India (ISRO), Italy (ASI), Japan, Russia (RKA) and the United States (NASA). The IADC is organized in four Working Groups (Measurements, Environment and Data Base, Protection and Mitigation, and a Steering Group. Periodic meetings of IADC result in the creation of new action items on the various issues of space debris hazard assessment and mitigation. Four such action items were adopted in Houston in 1997 at the 15th meeting of the IADC dealing with LEO constellation modeling, reentry survivability, preparation of a hypervelocity impact protection manual, and hypervelocity impact test facility calibration. In addition, on-going cooperation in LEO and GEO debris observation campaigns, a compilation of orbital debris sources, and the exchange of information on risk objects nearing reentry was continued at its 16th meeting at Toulouse, France on November 3-6, 1998.

c) United Nations (UN)

A Committee on the Peaceful Uses of Outer Space (UNCOPUOS) represented by its Scientific and Technical subcommittee held its thirty-fifth session in Vienna February 9-20, 1998. The representatives from fifty-four countries including the United States attended the session. Also present were representatives of the European Space Agency (ESA), committee on Space Research (COSPAR), International Academy of Astronautics (IAA), International Astronautical Federation (IAF), International Astronomical Union (IAU), International Society for Photogrammetry and Remote Sensing (SPRS) and International Space University (ISU).

General matters related to space debris were discussed at the meeting where it was agreed that consideration of space debris was important and that international cooperation was needed to expand appropriate and affordable strategies to minimize the potential impact of space debris on future space missions. The subcommittee further agreed that national research on space debris should continue and that Member States and international organizations should make available to all interested parties the results of such research including information on practices adopted that had proved effective in minimizing the creation of space debris.

A technical report on space debris was adopted by the Scientific and Technical subcommittee of UNCOPOS (United Nations Techn. Rpt., 1999).

1.5. Summary and Conclusions

This chapter reviewed the nature and the magnitude of the space debris hazard at present and in the near future to operational space systems. It can be concluded that the threat that orbital debris poses to space activities is currently not large but is anticipated to increase in time if appropriate mitigation measures are not implemented. The cost and effort of taking judicious and timely steps now to enhance the understanding of the risk and agree on ways to reduce it can be relatively small. Risks may increase significantly if no action is taken now.

CHAPTER 2

EUROPEAN SPACE AGENCY ACTIVITIES ON ORBITAL DEBRIS

W. Flury

The European Space Agency's (ESA) objectives in the field of space debris have been defined by its Council in 1989. ESA's debris related activities comprise research, application of debris reduction measures, and international cooperation, both on a European and world-wide level. The research activities address the knowledge of the terrestrial particulate environment, risk assessment of manned and unmanned space missions, and protective and preventive measures. An overview will be given of major debris research activities of ESA including the DISCOS database of space objects, the MASTER reference model for the space debris and meteoroid environment, the ESABASE/DEBRIS analysis tool for predicting impacts and resulting damage on a 3D spacecraft geometry, shield design aspects and risk assessment for the ESA module of the International Space Station (ISS), and debris mitigation and risk reduction measures. Since space debris is a worldwide problem ESA pays great attention to international cooperation. ESA is a founding member of the Inter-Agency Space Debris Coordination Committee (IADC). ESA is actively participating in the deliberations on space debris in the Scientific and Technical Subcommittee of the United Nations Committee on the Peaceful Uses of Outer Space (UNCOPUOS).

2.1. Introduction

Already in the late seventies ESA has addressed space debris issues, e.g. the collision risk in the geostationary orbit and the re-entry of risk objects. Following the issue of the *ESA Report on Space Debris* (Space Debris — The Report of the ESA Space Debris Working Group, 1988), in 1989 the Council of ESA adopted a resolution on Space Debris (ESA/ C/LXXXVII/Res. 3 (final) "Resolution on the Agency's policy vis-à-vis the Space Debris issue") and approved ESA/C(89)24, rev. 1 "ESA Activities for Space Debris", where the Agency's objectives in the field of space debris were formulated:

- to minimise the creation of space debris to ensure free access to space and reduce the risk for manned and unmanned space flight;
- to reduce the risk on ground due to re-entry of space objects;
- to reduce the risk for geostationary satellites;
- to acquire through its own facilities and in cooperation with other space agencies the data on space debris which are necessary to assess the extent of the problem and its consequences;
- to study the legal aspects of space debris.

Despite reduced launch activities and application of debris control measures, the space debris population in orbit is steadily growing with a corresponding increase of the hazard. New aspects are introduced with multi-satellite constellations in low Earth orbit (LEO) and the increased use of micro-satellites, which present an increasing challenge for space debris mitigation. As a consequence the priorities of the Agency's space debris policy as adopted in 1989 will also have to be directed towards paying more attention to the near Earth orbit region with its more dense debris population. ESA's space debris activities are reviewed regularly by the Council and adapted to new developments.

This paper gives an overview of ESA's debris related activities, which comprise research, application of debris reduction measures, and international cooperation, both, on a European and worldwide level. Most research activities are carried out in ESA's member states by specialized groups in academia and industry (Flury (ed.), 1993; Kaldeich and Harris (eds.), 1997; Flury (ed.), 1995; Flury (ed.), 1997; Flury and Klinkrad (eds.), 1999).

2.2. Space Debris Research Activities

2.2.1. The Terrestrial Meteoroid and Debris Environment

The objective is to gain a more comprehensive knowledge of the space debris population in terms of size and spatial distribution and to upgrade the European capabilities for obtaining the necessary observations. Main areas for research are the altitude band between 300 km and 1600 km, the geotransfer space at low inclination, and the geostationary orbit.

The Agency has established the DISCOS space debris database at ESOC for its own use and entities in the Member States (Jehn, Vinals-Larruga and Klinkrad, 1993). It supplies information on the currently catalogued objects and is a basis for risk assessments and understanding of the evolution of the debris environment. Among other data DISCOS contains currently some 3'200'000 records of the NASA Two-Line Elements (TLE) which are orbital element sets of all known (about 8'700) catalogued space objects[1]. Upgrading of the DISCOS debris database is ongoing.

---

[1] The minimum size of the catalogued (or trackable) objects is about 10–50 cm in LEO and about 1 m in the geostationary orbit (GEO). They are tracked by radar and optical sensors of the United States' Space Command.

Further to the DISCOS database a comprehensive ESA meteoroid and debris reference model (MASTER[2] model) has been completed (Klinkrad, Flury *et al.*, 1994; Flury, 1995; Klinkrad, 1996; Klinkrad, Bendisch *et al.*, 1997; Wegener *et al.*, 1999). The model, which became available for general distribution on a CD-ROM in 1997, describes the debris and meteoroid environment from LEO to the geostationary orbit (GEO) for a minimum objects size of 0.1 mm. The model is based on the catalogued population and on the known breakups of spacecraft and rocket upper stages in orbit. It constitutes a common basis in Europe for debris-related analyses in industry, research institutes and space agencies. Validation and upgrading of ESA's MASTER model for space debris are ongoing. Objectives of the upgrade are to include particles as small as 1 micrometer, and non-fragmentation debris (Wegener *et al.*, 1999). The meteoroid model has been upgraded by the Max-Planck-Institute for Nuclear Physics, Heidelberg (Gruen, Zook *et al.*, 1985; Staubach and Gruen, 1996). Recent dust measurements of interplanetary probes (Galileo and Ulysses) have been taken into account. Methods for the calculation of the long-term evolution of Earth orbiting debris (up to 100 years) have been developed (Anselmo, Cordelli *et al.*, 1996; Bendisch and D. Rex, 1996).

Significant contributions have been made to the knowledge of small-size (micron to millimeter) particles with EURECA and the returned solar array of the Hubble space telescope (HST) (Drolshagen, McDonnell *et al.*, 1996; McDonnell, Griffiths *et al.*, 1995). All outer surfaces were systematically surveyed for impacts. Observed impact features range in size from about 3 microns to a maximum of 7 mm of shattered cover glass. Thousands of craters were visible with the naked eye. About 150 impacts have completely penetrated the solar array of the HST. The observed number of impacts agrees largely with the models for craters smaller than about 1 mm and exceeds the predictions for the larger craters. In several European institutes samples from EURECA and the HST solar array have been analyzed for impact residues and hypervelocity calibration tests have been performed. The EURECA and HST impact data are available in an electronic database.

Experiments on micro-debris and cosmic dust are also carried out on the MIR space station, e.g. ESEF on EUROMIR, a flight instrument attached to exterior of MIR (Maag *et al.*, 1997). The instrument was designed to measure the impacts and trajectory of micro-debris and meteoroids. The spare cosmic dust detector of the Ulysses mission has been placed on the Russian geostationary EXPRESS-2 spacecraft (launch 26 September 1996). This allows monitoring the micro-debris particles in the geostationary orbit (GORID experiment, Drolshagen *et al.*, 1999). The interpretation of the measurements is, however, not a straightforward process due to the interaction with charged particles in the Earth's environment.

Efforts are made to improve the capabilities of small-size particle detectors in order to gain more accurate information on relevant parameters (e.g. impact velocity and mass). The standard in-situ detector DEBIE for small-size particles is in development. Its first application will be on the PROBA spacecraft which will be placed in a near-polar circular orbit near 1000 km altitude.

Current space debris models in LEO suffer from significant uncertainties for objects smaller than about 50 cm. This is of particular concern for spacecraft which require protection, as shielding against objects larger than 1 cm is technically not feasible.

---

[2] Meteoroid And Space Debris Terrestrial Environment Reference Mode

Therefore, a study was conducted to investigate the feasibility of detection and tracking of mid-size debris (1–50 cm) with the radar facilities of FGAN (Forschungsgesellschaft für Angewandte Naturwissenschaften), located at Wachtberg (Germany). FGAN operates a high power radar system TIRA (tracking and imaging radar) which is able to track and image aircraft, functioning satellites and space debris (Mehrholz, 1995). It is composed of three main sub-systems: a 34-meter parabolic dish antenna, a L-band tracking radar (1.333 GHz, wavelength 22.5 cm), and a Ku-band imaging radar (16.7 GHz, wavelength 1,8 cm). Test campaigns have been conducted with the ODERACS experiment (Orbital Debris Radar Calibration Spheres of 5 cm, 10 cm, 15 cm diameter), launched from the US Space Shuttle in February 1994 into orbits near 350 km altitude.

A 24 hours measurement campaign in low Earth orbit was carried out on Dec. 13–14, 1994 with the FGAN radar facilities. The purpose of the campaign was to detect in a specific volume near 800 km altitude all objects above a minimum size, and compare the detections with those predicted by a space debris model. The pointing direction of the radar remained unchanged (beam park experiment). The minimum size of detectable objects with the FGAN radar at that altitude is near 2 cm. Post-processing of the data (300 Gbyte) shows 255 detections. 72 of them are catalogued objects. The MASTER model underpredicts the detections by a factor of 2. For the observations of the Haystack radar (USA) the MASTER model overpredicts slightly the number of detections. Such measurement campaigns are excellent methods for the validation of the MASTER model in the several centimeters size range.

An even more powerful experiment has been carried out in November 1996: COBEAM-1/96, a bi-static radar experiment with FGAN transmitting and receiving and the world's largest steerable radio telescope at Effelsberg receiving. It allowed to collect information of centimeter-size objects in LEO (Mehrholz, Leushacke and Jehn, 1999; Leushacke, Mehrholz and Jehn, 1997).

FGAN carried out additional 24-hours beam park experiments in February and April 1999. The number of detections was 378 in Feb. 1999. It includes 69 cataloged objects. From the 352 detections in April 1999 the 83 were cataloged objects.

A 1-meter aperture Zeiss telescope is being installed at the Teide Observatory in Tenerife. The telescope will serve primarily as check-out facility for the optical data link between ARTEMIS and SPOT-4 (SILEX experiment). The large field of view optical system dedicated to space debris investigations will be equipped with a 4096 × 4096 pixel CCD camera. The minimum size of detectable objects is expected to be 2–6 cm in LEO and 20–40 cm in GEO. With this telescope which is planned to become operational in 2000, European space debris observation capabilities in geostationary transfer orbits and the geostationary orbit will be substantially improved.

It should be noted, however, that in the area of space surveillance and knowledge of the environment, Europe is heavily dependent on external sources. European ground-based observation capabilities (radar, optical) can be improved through optimum use of national facilities (e.g. radar and optical facilities in France, Germany, Switzerland and United Kingdom) and ESA facilities (e.g. 1-Meter Zeiss telescope on the Teide observatory).

Another research area is space-based detection. Studies have shown that with small aperture (10 cm) optical instruments valuable information can be gained on the mid-size debris population (1 to 50 cm).

2.2.2. Risk Analysis

An example of the growing hazard in space is the situation of ESA's Earth observation satellites ERS-1 and ERS-2, which are in sun-synchronous orbits at 770 km altitude. During a five-years period the chance of an ERS satellite to collide with a 1 cm size particle — which could lead to severe damage or destruction — amounts to 1 to 2%. A proximity analysis with current data from the DISCOS database has shown, that during a period of a few days several near-misses (typical minimum distance 0.3 to 2 km) between ERS and other catalogued space objects (satellites, rocket, stages, large fragments) occur. A collision with a large object would very likely be fatal for ERS because of the large impact velocity (on average near 13 km/s). For ERS-1 & ERS-2 proximity predictions are being carried out routinely.

ESABASE/DEBRIS is ESTEC's 3D software tool for the impact risk analysis (number of impacts and damage assessment) of manned and unmanned spacecraft due to debris and meteoroids. For potential damage assessment and post-flight analysis ESABASE/DEBRIS has been applied to various projects, e.g. EURECA, HST, Columbus COF, ENVISAT, ISO and XMM. ESABASE/DEBRIS will be further upgraded for application to the Agency's programs. This includes implementing new damage laws, new results in material response under impact, and advanced methods for risk assessment.

As space debris may endanger the function or the survival of large space vehicles, detailed risk assessments have been conducted for the Columbus Orbiting Facility. The overall Space Station protection requirement against meteoroids and debris asks for a probability of at least 81% over 10 years of operation of no penetration of critical items such as propellant tanks and manned modules.

2.2.3. Debris Protection and Mitigation

The flux of space debris in LEO has reached such a level that shielding of manned vehicles is needed. The growing amount of man-made objects in LEO may, however, also require the protection of sensitive parts of unmanned spacecraft, e.g. the electronic boxes of Earth observation spacecraft. In order to understand how to protect against space debris, first the effect of debris on spacecraft structures must be investigated. Generic work on hypervelocity impact tests and protective measures for manned vehicles (shield design) have been carried out at several facilities within the Member States (Lambert, 1993; Stilp, 1997). Upgrading of impact test facilities and numerical methods is necessary for realistic simulations of the space environment and its effect on space systems. Experimental tests and numerical simulations should extend on impact direction and velocity, pressure regime, and type of materials.

The ESA technological activities in the field of shielding against space debris are divided into three domains:

- characterization of material
- development of advanced test techniques
- validation of computer codes for hypervelocity impact simulation.

Characterization of materials covers the determination of impact damage laws for insufficiently documented materials or configurations. Carbon fiber reinforced plastics (cfrp), multi-layer insulations and sandwich panels with cfrp facing and aluminium honeycomb cores have been tested. New materials for re-entry vehicles are studied: ablative materials such as AQ60 developed by Aerospatiale, carbon-carbon, carbon-sic and flexible external insulations which are made of silica felt and Nextel.

Experimental work is being performed to study hypervelocity impacts on pressurized vessels. Boundaries between simple impact holes and catastrophic burst are being determined for gas filled pressure vessels made of aluminum and titanium.

The experimental simulation of impact conditions expected in LEO requires test techniques able to accelerate projectiles at least up to velocities around 11 km/s. Two techniques have been developed. Shaped charges are used to generate cylindrical projectiles, while multi-stage active disk launchers accelerate flat disks up to the required velocity. These advanced techniques are mostly used to obtain experimental impact data on state of the art shields. These data are compared with computer codes impact simulations in order to validate selected codes.

In the shielding area research activities have focused on the selection of optimum materials and the design of shields with minimum mass for a given safety and damage tolerance. The objective is to reach cost-effective protection of unmanned spacecraft in LEO and to design shield and protection of manned vehicles, e.g. the planned ESA module of the International Space Station, in compliance with the safety requirements of the Agency.

The Columbus shielding performances have been quantified experimentally on limited size unpressurized flat samples in the velocity ranged between 3 and 7 km/s. The baseline configuration (double bumper) limits efficiently the damage of the pressurized module for particles below 0.5–1.2 cm size.

The debris preventative measure of venting the Ariane upper stage has been carried out routinely from flight 59 (V59, Sept. 1993) onwards, regardless of the type of the target orbit. Debris mitigation measures for Ariane V (passivation of the upper stage) have been implemented (Sanchez, Bonnal and Naumann, 1997).

First steps for the reduction of mission-related objects are being applied by projects. During the operational lifetime of ENVISAT the creation of mission-related objects will be precluded. At end-of-life controlled venting of pressure vessels and residual fuel, discharge of batteries, and shutdown of the power system are foreseen.

Several geostationary satellites have been reorbited into a graveyard orbit above the geostationary orbit (the figures in bracket indicate the achieved altitude increase and the year of reorbiting, Janin, 1999): GEOS-2 (260 km, 1984), OTS-2 (318 km, 1991), METEOSAT 2 (334 km, 1991), ECS-2 (335 km, 1993), OLYMPUS (-213 km, 1993), METEOSAT 3 (940, 1995), METEOSAT 4 (833 km, 1995), MARECS-A (1536 km, 1996), ECS-1(377 km, 1996). ESA's policy is to reorbit into a graveyard orbit with a perigee located at least 300 km above the geostationary orbit. Due to a spacecraft failure OLYMPUS could not be inserted into a graveyard orbit above the geostationary ring. The approach taken by ESA has been confirmed by the adoption by ITU of the recommendation ITU-R S.1003 on the environmental protection of the geostationary orbit.

At the occasion of the uncontrolled re-entry of the equipment module of the Chinese spacecraft 1993-063H in October 1993, ESOC provided forecasts to ESA's Member States on time and location of the re-entry based on data from European facilities, US Space Command, and Russia. Important radar data were obtained from the radar facility of FGAN at Wachtberg-Werthhoven (Germany). The experience gained and cooperation established during this campaign will be helpful to cope with similar events in the future. The re-entry module 1993-063A reentered in March 1996 in the South Atlantic Ocean. As a consequence of the thermal protection system, it can be expected that the spacecraft did only partially burn up during the atmosphere entry.

ESOC provided also information to its Member States at the time of the uncontrolled re-entry of Kosmos 398 in December 1995. Kosmos 398, a spacecraft with a mass of several tons, has been launched in 1971 as part of the Soviet Lunar Programme.

Currently, the growth of the space debris population is determined by launch operations and in-orbit breakups (explosions). Accidental collisions are not yet a growth factor. However, current practices could lead to a self-sustaining proliferation process as a consequence of collisions among large objects. More efficient mitigation measures have to be considered, which may include removal from space of rocket bodies and spacecraft after completion of their mission (Flury, Heusmann and Naumann, 1995). Therefore, strategies for de-orbiting of spacecraft and rocket upper stages in LEO have to be elaborated. Such measures will have far-reaching implications with regard to design and cost of space system. Thorough cost-benefit trade-offs will be needed to identify the suitable disposal of decommissioned spacecraft and rocket stages. The issue of disposal orbits must be carefully examined. Reorbiting should only be considered as an interim measure, since useful spatial regions will not be accessible. Ultimately, removal from orbit will be required.

ESA has defined in 1988, as part of the safety policy, a specific requirement for prevention of debris creation in PSS-01-40. The PSS documents, which represent formal standards for space system design, are in the process of being replaced by the European Cooperation for Space Standardisation (ECSS). Suitable standards addressing the space debris issue have to be introduced in ECSS for design and operation of space systems.

In 1999 the ESA Debris Mitigation Handbook (Klinkrad, 1999) has been issued. The Handbook introduces the space debris problem to designers and operators and advises on debris mitigation concepts. It allows to assess the hazards to spacecraft caused by space debris and meteoroids. The Handbook has no regulative character.

2.3. Harmonization in Europe and International Cooperation

As space debris is a global problem which only can be solved by a joint effort, discussions and cooperation with Member States, other space agencies and relevant organizations are of great importance.

Harmonization has proceeded at European level between Member States and world-wide cooperation between ESA and other space agencies within the framework of the Inter-Agency Space Debris Coordination Committee.

Coordination meetings are being held regularly with the national space agencies ASI (Italy), BNSC (United Kingdom), CNES (France) and DLR (Germany). Main issues are the harmonization of European space debris research activities and the coordinated use of national facilities for space debris observation (radar, optical), hypervelocity impact tests, analysis of material returned from space and the identification of methods and approaches for debris reduction. The research programmes are closely coordinated between the national Space Agencies and ESA such as to avoid duplication and reduce overlapping of activities.

The Inter-Agency Space Debris Coordination Committee (IADC) with its current members ESA, Japan, NASA, RKA (Russia), ASI (Italy), BNSC (United Kingdom), CNES (France), CNSA (China), DLR (Germany), and ISRO (India), offers a forum for discussion and coordination of technical space debris issues. The primary purpose of the IADC is to exchange information on space debris research activities, to facilitate opportunities for cooperation in space debris research, to review the progress of on-going cooperative activities and to identify debris mitigation options. IADC comprises a Steering Group and four technical Working Groups dedicated to specific areas: I) measurements of the environment, II) database and environment, III) protection, iv) mitigation. Within the framework of this cooperation the participants exchange relevant technical information and experience related to space debris and prepare common strategies to counter the space debris problem. Current activities focus on

- a joint debris database,
- reentry of risk objects,
- improved meteoroid and debris models,
- debris mitigation in LEO,
- optical observations and debris detection in the geostationary ring,
- debris management practices in the geostationary ring.

IADC is instrumental in providing the technical know-how and the technical and financial feasible strategies for debris control.

As an active member of IADC, ESA is contributing to cooperative activities, such as the establishment of a common database, the execution of common observation campaigns, and the identification of mitigation measures.

Since 1994, the topic *space debris* has been on the agenda of the Scientific and Technical Subcommittee of the United Nations Committee on the Peaceful Uses of Outer Space (UNCOPUOS). The S&T Subcommittee adopted in 1995 a multi-year work plan:

- 1996: Measurements of space debris and effects of the environment on space systems
- 1997: Modeling of the space debris environment and risk assessment
- 1998: Space Debris mitigation measures (protection, prevention, removal)
- 1999: Consolidation and completion of the UN Report on Space Debris.

ESA is supporting the technical discussions at the Scientific and Technical Subcommittee of UNCOPUOS. ESA has contributed to the multi-year work plan 1996–1999, which addresses all major technical aspects of space debris.

2.4. Conclusions

ESA has recognized the serious nature of the space debris problem since many years. ESA has adopted a systematic approach towards the problem, which is based on research, application of mitigation measures, and international cooperation.

Current risk levels due to space debris are low but are increasing.

Despite the application of debris reduction measures number and mass of debris objects are increasing. Expendable space transportation systems are a major debris source. Reusable launch systems would eliminate this source.

New developments such as multi-satellite constellations could strongly influence the space debris environment.

Ultimately more efficient debris control measures will be needed, such as selective removal from orbit of used upper stages and decommissioned spacecraft. A critical issue is the use of graveyard orbits.

In order to deal with the space debris problem in an effective manner, cooperation among all space-faring nations is required. The IADC is of paramount importance for the identification of cost-efficient debris mitigation measures.

In order to ensure that the hazards to space operations caused by space debris will remain within acceptable limits, ultimately, a code of conduct or a legal instrument established through the United Nations will be needed.

CHAPTER 3

A MATHEMATICAL MODEL FOR SPACE DEBRIS EVOLUTION, PRODUCTION AND SELF-PRODUCTION*

N.N. Smirnov

Investigations and long-term forecasts of orbital debris environment evolution in low Earth orbits are essential for future space missions hazards evaluations and for adopting rational space policies and mitigation measures. The paper introduces a new approach to space debris evolution mathematical modeling based on continua mechanics incorporating differential equations in partial derivatives. This approach is an alternative to the traditional approaches of celestial mechanics incorporating ordinary differential equations for model fragments. The continua approach to orbital debris evolution modeling has big advantages for description of the evolution of fraction of particles numbering billions, because it replaces the traditional tracking of space objects by modeling the evolution of their density of distribution.

3.1. Introduction

The number of trackable debris (larger than 10 cm^2) has been increasing since the beginning of the space era and now reaches 8 700 objects with a total mass of $2.3 \cdot 10^6$ kg. The smaller fraction (1 + 10 cm) incorporates fragments generated as operational debris (separating construction elements) and as a result of orbital explosions and collisions. Their number is estimated as 20 000 pieces (that is 0.5% of the total amount of all the technogeneous

---
* The present research was supported by the Russian Foundation for Basic Research (projects 98-01-00218 and 00-15-99060) and Program "Fundamental Research of High School, Universities of Russia" (project 99-2153).

fragments) with a total mass of about 10^3 kg. The number of the particles of $0.1 \div 1$ cm originated as results of orbital explosions and collisions is estimated as $3.5 \cdot 10^6$ objects. The last statistical estimates made in (OSTP, 1995) give an order of magnitude higher values for the number of untrackable objects: $1.1 \cdot 10^5$ of objects sizing within the range of $1 \div 10$ cm, 35 bln. of objects sizing $0.1 \div 1$ cm. All those objects remain in orbit for a sufficiently long time to become a hazard to space activities (Chobotov, 1990; Smirnov et al., 1993). The number of smaller objects in orbits could be hardly evaluated. The space-based active and passive detectors of the orbital debris and meteoroid environment gave the possibility to evaluate the cumulative flux of the particles in the range of 10^{-4} cm $\div$ 10^{-1} cm. The uncertainty of the estimates of the cumulative flux of small particles of a micron size is very large: $3 \cdot 10^3 \div 2 \cdot 10^5$ objects/($m^2 \cdot$ year), that gives the discrepancy of two orders of magnitude (UN Technical Report on Space Debris, 1999). Thus additional measurements should be performed detecting the cumulative flux of particles of the range more precisely.

The methods for investigating the space debris evolution are now being developed rather rapidly (Reynolds, 1991; Krisko, 1999; Bendisch et al., 1999; Walker et al., 1999). The most widespread approach to space debris production and evolution forecasts (models EVOLVE, MASTER and IDES) is based on tracking all the space objects using traditional methods of celestial mechanics (Reynolds, 1991). The sources of contamination are being modelled stochastically. The results are averaged after a sufficient number of realisations. The disadvantages of such an approach are evident: the method is too costly and time consuming even for simplified statistically averaged contamination models. Introducing into traditional models more realistic break-up and fragmentation models each being time consuming as well will make long-term forecasts practically impossible because the time required for numerics will grow exponentially along with the number of collisions on approaching the cascade process.

Thus the scientific community is now faced the two alternative possibilities: either to continue getting hold of the existing approach in modeling using simplified break-ups and fragmentation description, or to give up the approach of tracking individual space objects turning to the description of their spatial distribution function evolution within the frames of continuous models accounting for the peculiarities of different types of orbital break-ups.

The traditional approach is very effective for the description of the orbital debris environment evolution at the early stages when the number of collisions and break-ups is relatively small. The break-ups description incorporated into the EVOLVE model had essential deviations from the results of real break-ups observations (Potter & AnzMeador, 1994, 1996) in terms of cumulative flux of large fragments. That could be the reason for essential deviations in forecasts of the beginning of the cascade process of the debris self-production. The forecasts made under similar assumptions for human space activity gave the estimates for the time left before the irreversible debris self-production could start, ranging from 25–50 years (Kessler, 1991; Rossi, Cordelli et al., 1993; Rex, Eichler, 1993) to 250–300 years (Smirnov et al., 1993). The divergence of the order of magnitude does not seem satisfactory that makes us pay more attention to peculiarities of different types of orbital break-ups being the essential factor for the long-term forecasts.

The results of orbital collisions differ strongly depending on a number of factors: the sizes of colliding objects, their structure, mass, shape, relative velocity, etc. Thus, the known

long-term forecasts of the cascade effect are based on the essential simplifications of the
conditions and the results of orbital collisions. For example, some forecasts are obtained
taking into consideration collisions of large (>10 cm) catalogued objects only. The frequency
of such collisions nowadays is evaluated as one per 30 years. At the same time, the probability
of collisions of small objects not taken into account is evaluated to be several orders of
magnitude higher. Thus, the problem of long-term forecasting the orbital debris evolution
accounting for the results of collisions (the cascade effect) is far from being solved.

The suggested in the present paper theoretical approach to modeling the long-term
debris evolution with account of a variety of break-ups in collisions and internal explosions
allows to overcome the enormous difficulties encountered while using the traditional
approach. The suggested approach abandons the traditional tracking of space objects
replacing it with modeling the evolution of the density of distribution of all the objects
(trackable and untrackable ones) with account of variety of collision scenario within the
frames of the approach of the continua mechanics. The cost we need to pay for this unique
possibility given by the continual approach is to develop adequate models describing the
collisions probability, variety of break-up scenario, the influence of the self-cleaning of
the low Earth orbits and the impact of new launches to be incorporated into the source
terms of the evolution equations for the continua. That would provide the possibility of
predicting and comparing the orbital debris long-term evolution forecasts for different
scenario of human space activity.

The predominant point of view among the debris community is that collisions usually
contribute to the growth of the orbital debris population. The preliminary estimates of the
collision probabilities show that the collision probabilities for small fragments to collide large
fragments are several orders of magnitude higher than that for the large catalogued objects.
Thus, taking into account fragmentation in collisions of small untraceable objects is essential
for adequate description of the cascade debris self-production effect. Besides, collisions of
small objects bring to formation of large amount of fine particles with velocities much less
than that of initial colliding fragments. Those particles will sediment more rapidly being
entrapped by the atmosphere contributing to a self-cleaning of the low Earth orbits.

Thus, collisions of different types of fragments could contribute both to the debris self-
production and to self-cleaning effects. Adequate description of the collision probabilities
matrix with spatially dependent elements is essential for long-term forecasts of the orbital
debris evolution.

3.2. Mathematical Model

An innovative approach to orbital debris evolution modeling is based on the continua
mechanics methodology. The description of the evolution of artificial objects density
distribution will be performed with account of possible collisions and their consequences.
To describe the evolution of the fields of concentration of debris fragments in space and
their velocities, the governing set of the evolution equations in partial derivatives will be
constructed with the source terms accounting for the multiplicity of space objects collision
consequences.

Thus the traditional approach of celestial mechanics to the orbital debris evolution modeling incorporating ordinary differential equations for each representative particle will be changed for the novel approach of continua mechanics incorporating differential equations in partial derivatives. The suggested approach has an analogy with methodizes of investigation of the rarefied flows in the upper atmosphere. In both cases the concentration fields, and not individual particles, are investigated taking into account collisions. The major differences for the cases are in different laws for particles motion, their different physical nature, different sources for the particles production and different collision consequences.

Essential differences in debris particles sizes and thus collision consequences brings to a necessity of introducing a number of "phases" or "mutually penetrating continua" into the model, each phase being characterized by its own density of distribution $\bar{\rho}_j$. The particles could be assembled into groups ("phases") due to the following attributes: their characteristic size d_j; the perigee altitude of the orbit $h_{\rho j}$, eccentricity e_j; inclination of the orbit i_j; ballistic coefficient. Let us assume that the whole number of debris particles could be grouped into N different phases. Then the mass conservation law results in the following mass equation (Smirnov *et al.*, 1993):

$$\frac{\partial \bar{\rho}_j}{\partial t} + div\bar{\rho}_j\vec{v}_j = I_j; \; j=1,...,N_p \tag{3.1}$$

where $\vec{v}_j$ is the local velocity of the j-th phase, $\bar{\rho}_j$ — distributed mass density of debris particles of the j-th phase. Along with the distributed mass density one could introduce a distributed volume density α_j of debris objects per volume unit of space, and the number density n_j of objects per volume unit. Then the following relationship between the introduced characteristics is valid:

$$\bar{\rho}_j = \alpha_j\rho_j \equiv n_j\frac{\pi d_j^3}{6}\rho_j, \tag{3.2a}$$

where ρ_j is the actual density of the material of debris particles, d_j — the effective characteristic diameter. The relationship (3.2a) presumes a compact shape of fragments (nearly spherical); for flat fragments originating in thin-walled shells breakups, one has a different relationship between the density and number of particles:

$$\bar{\rho}_j = \alpha_j\rho_j = n_j\pi d_j^2\delta\rho_j, \tag{3.2b}$$

where δ is the thickness of shell elements.

The term I_j in the equation (3.1) characterizes the mass exchange of the j-th phase with other phases and mass variations of the j-th phase due to increase of operational debris and fragmentation of large space objects and due to elimination of the particles which size turns to be below the critical one. In breakup of a debris particle of the j-th phase the fragments can join other phases (as their size turns to be smaller than that of the initial particle). The sizes of some fragments could turn to be below the minimal size of particles, taken into consideration by the model. Those particles are either not hazardous to space flights or slow down in the atmosphere fast enough not contributing to long-term evolution scenario. Based on the analysis one could suggest the following structure for the mass exchange term in the equation (3.1):

$$I_j = \sum_{k=1}^{N} \kappa_{jk} + M_{j\,op} + M_{j\,ex} - \mu_j,$$ (3.3)

where κ_{jk} is the mass flux from the k-th to the j-th phase due to fragmentation of debris particles in collisions; $M_{j\,op}$ is the mass contribution to the j-th phase from the operational debris occurring in realization of new space programs; $M_{j\,ex}$ is the mass flux to the j-th phase due to destruction of large space objects (orbital explosions of the upper stages of rockets, etc.); μ_j is the mass decrease of the j-th phase, that does not contribute to the mass increase of other phases.

The momentum equation for phases within the frames of the continua mechanics approach has the following form

$$\frac{\partial \bar{\rho}_j \vec{v}_j}{\partial t} + div \bar{\rho}_j \vec{v}_j \otimes \vec{v}_j = \vec{F}_j + \vec{F}_{dj} + \vec{P}_j + \vec{K}_j,$$ (3.4)

where $\vec{F}_j$ characterizes the mass forces, $\vec{F}_{dj}$ — the influence of the atmospheric drag, $\vec{P}_j$ — the pressure of Sun radiation, $\vec{K}_j$ — the momentum flux to the j-th phase due to the mass exchange I_j:

$$\vec{K}_j = \sum_{k=1}^{N_p} \kappa_{jk} \vec{v}_{jk} + M_{j\,op} \vec{v}_j + M_{j\,ex} \vec{v}_j - \mu_j \vec{v}_j,$$ (3.5)

where $\vec{v}_{jk}$ is the velocity of the particles having come to the j-th phase from the k-th one. For positive values of κ_{jk} those velocities are normally close to $\vec{v}_j$, else those particles would not stay in orbit. For negative values of κ_{jk} particles, joining the k-th phase should have velocities close to $\vec{v}_k$. Nevertheless, deviations are possible, but the following condition of agreement should be satisfied:

$$\sum_{k=1}^{N_p} \sum_{j=1}^{N_p} \kappa_{jk} \vec{v}_{jk} = 0.$$ (3.6)

The mean volumetric mass force present in (3.4) could be determined by the formula

$$\vec{F}_j = -\bar{\rho}_j g(\bar{x}) \vec{e}_r,$$ (3.7)

where $g(\bar{x})$ is the gravity acceleration, $\vec{e}_r$ — the physical component of the radial basis vector in a spherical system of coordinates. The atmospheric drag term in the equation

(3.4) could be defined in the following way

$$\vec{F}_{dj} = -\frac{1}{2}c_f^j \rho_a(\vec{x},t)\vec{v}_j |v_j| \frac{3\alpha_j}{2d_j},$$
(3.8)

where $\rho_a(\vec{x},t)$ is a spatial density distribution in the upper atmosphere, that strongly depends on altitude. Temporal density variations should be also taken into account. The 11-year cycle of the Sun activity causes the density variations in the upper layers of the atmosphere up to 20 times, that produces a serious effect on debris fragments sedimentation (Kasimenko, Ryhlova, 1995). The equation (3.8) contains the drag coefficient c_f^j for the j-th phase particles moving in a rarefied gas. For spherical particles this coefficient could be determined using the following formula:

$$c_f^j = \frac{2e^{-\beta_j^2}}{\sqrt{\pi}\beta_j^3}(2\beta_j^2 + 1) + \frac{erf(\beta_j)}{\beta_j^4}(4\beta_j^4 + 4\beta_j^2 - 1) + \frac{4k_{ac}\sqrt{\pi}}{\beta_{jw}},$$
(3.9)

where

$$\beta_j = v_j\sqrt{\frac{m_a}{2kT_a}}, \beta_{jw} = v_j\sqrt{\frac{m_j}{2kT_j}}, erf(\beta_j) = \frac{2}{\sqrt{\pi}}\int_0^{\beta_j} e^{-x^2}dx;$$

m_a — the mean molar mass of gases in the upper layers of the atmosphere; k — the Boltsman constant, T_a — gas temperature, T_j — particles surface temperature; k_{ac} — accommodation coefficient taking into account the character of gas molecules reflection from the surface of a moving particle.

The motion of small particles in Space is influenced essentially by the pressure of Sun radiation, that should be taken into account for adequate long-term evolution modeling (Poljakhova, 1995; Mikisha, Smirnov, 1995). There exist two types of solar radiation: light (photons) emission propagating at a speed of $c = 3 \cdot 10^8$ m/s and emission of corpuscles (solar wind) propagating at a speed of $w = 4 \cdot 10^5$ m/s. Both types of the radiation exert pressure on space objects and cause aberration dynamic effects. Thus the radiation pressure term $\vec{P}_j$ present in the equation (3.4) is actually the sum of four effects:

– direct photon radiation pressure p_r,
– direct solar wind pressure p_r^b,
– aberration dynamics Poyting-Robertson effect of photon radiation $\cong p_r \frac{\vec{v}_j^H}{c}$,
– aberration dynamics effect of a solar wind $\cong p_r^b \frac{\vec{v}_j^H}{\omega}$,

where $\vec{v}_j^H$ is a heliocentric velocity of particles;

$$p_r = \frac{q_r}{R_H^2}; \ p_r^b = \frac{q_r^b}{R_H^2};$$
(3.10)

R_H is the distance from the Sun; $q_r = 0.101 \cdot 10^{18}$ kg·m/s², $q_r^b = 0.23 \cdot 10^{14}$ kg·m/s² — intensities of photons and corpuscles emissions by the Sun respectively.

Besides debris particles are essentially affected by the sun radiation only being out of the Earth shadow. Thus the mean volumetric force $\vec{P}_j$ in the equation (3.4) could be determined by the following approximate formula:

$$\vec{P}_j = n_j \frac{\pi d_j^2}{4} H(-\vec{R}_H \cdot \vec{R}) \left[(p_r + p_r^b) \frac{\vec{R}_H}{|R_H|} - \left(\frac{p_r}{c} + \frac{p_r^b}{\omega} \right) \vec{v}_j^H \right], \tag{3.11}$$

where $\vec{R}_H$, $\vec{R}$ — the radius-vectors of a particle in a heliocentric and a geocentric coordinate system respectively;

$$H(x) = \begin{cases} 0, & \text{if } x \le 0 \\ 1, & \text{if } x > 0 \end{cases}.$$

The Heavyside function $H(x)$ in the formula (3.11) is equal to zero for particles hidden in the Earth's shadow. The formula (3.11) could be transformed using (3.2):

$$\vec{P}_j = \frac{3}{2} \frac{\alpha_j}{d_j} H(-\vec{R}_H \cdot \vec{R}) \left[(p_r + p_r^b) \frac{\vec{R}_H}{|R_H|} - \left(\frac{p_r}{c} + \frac{p_r^b}{\omega} \right) \vec{v}_j^H \right]. \tag{3.12}$$

The formula (3.12) shows that for the debris phases containing smaller particles the mean volumetric force in the equation (3.4) grows up as $\frac{1}{d_j}$. Thus the radiation effect is more essential for smaller particles.

The components κ_{jk} present in I_j and $\vec{K}_j$ terms of the equation (3.1), (3.4) determine the mass fluxes from the k-th to j-th phase being the results of fragmentations of the j-th phase in collisions with other space objects. The components κ_{jk} form an antisymmetric tensor by definition:

$$\kappa_{jk} = -\kappa_{kj}, \ k \ne j; \ \kappa_{jj} = 0, \ k = j. \tag{3.13}$$

The values of the components of this tensor depend on probabilities of collision of particles of the k-th phase with other phases and particles of j-th phase with other phases as well. The mechanisms of dynamical interaction and breakup of colliding objects, and the fragments mass and velocity distributions are of major importance as those parameters make it possible to determine the number of particles contributed to the j-th phase among all the spectrum of particles formed as a result of fragmentation.

Collision probabilities for in orbit collisions could be evaluated and risk analysis could be performed based on two different approaches: a deterministic one and a stochastic one (Khutorovsky *et al.*, 1995). By now both approaches for collision risk assessment for moving space objects are being developed. However, the present model is based on a continua Eulerian description of debris clouds, and tracking of individual debris particles is not foreseen. Thus the terms κ_{ij} present in the equations characterize the events taking place in a definite zone of space, and not in a Lagrangian particle. That brings to a necessity of modifying the formulas derived for determining collision probabilities of a moving space object with other debris particles.

The estimated number of k-th phase particles colliding a j-th phase particle for the time interval dt could be expressed by the formula

$$dP_{jk}^L = S_{jk} n_k v_{jk}^{rel} dt, \qquad (3.14)$$

where v_{jk}^{rel} is the relative velocity modulus; $n_k(\bar{x}, t)$ — spatially distributed particles number density; $S_{jk} = 0.25\pi(d_j + d_k)^2$ — mean collision cross-section.

To determine the spatially distributed collision probability one needs to take into account the number of j-th particles coming to cross the area S_{jk} in a definite place in an Euler space within the same time interval:

$$dP_j^E = S_{jk} n_j v_j dt. \qquad (3.15)$$

Defining the probabilities of collisions per time unit as

$$p_{jk}^L = \frac{dP_{jk}^L}{dt} = S_{jk} n_j v_{jk}^{rel}$$

$$p_j^E = \frac{dP_j^E}{dt} = S_{jk} n_j v_j \qquad (3.16)$$

one could obtain the estimated number of collisions of j-th and k-th particles per time per volume unit (spatially distributed collisions rate):

$$p_{jk}^E = \frac{p_{jk}^L \cdot p_j^E}{V_j^E} = \frac{S_{jk}^2 n_j n_k v_j v_{jk}^{rel}}{S_{jk} v_j} = S_{jk} n_j n_k v_{jk}^{rel}. \qquad (3.17)$$

The spatially distributed probability of collision of an object belonging to the j-th phase with objects belonging to all other phases is the sum of probabilities

$$\tilde{P}_j^E = \sum_{k=1}^N p_{jk}^E. \qquad (3.18)$$

It is seen from the (3.17), (3.18) that spatially distributed collision probabilities depend on the number densities products, velocities and the collision cross-section area. The same result was obtained before (Smirnov et al., 1993) in direct derivations based on kinetic theory assumptions.

To determine the mass exchange between phases caused by collisions and breakups (3.13) one needs to introduce special functions K_{kl}^j characterizing the mass of particles of the j-th phase originating as a result of fragmentation in collisions of k-th and l-th particles. Those functions depend on collision velocities, mass, density, material structure, etc. The functions $K_{kl}^{j}(v_{jk}^{rel}, d_k, d_l, \rho_k, \rho_l, ...)$ are external parameters for the present problem. Those functions could be determined from the existing solutions of hypervelocity impact, breakup and explosion problems, obtained within the frames of a deterministic (Smirnov *et al.*, 1997, Smirnov *et al.*, 1994, Kiselev, 1995, Smirnov *et al.*, 1996) or stochastic (Matney, Theall, 1999; Chobotov, Spencer, 1991) approach. A more detailed description of orbital breakups modeling peculiarities will be given in the Chapter 7 of the present edition.

The functions K_{kl}^{j} due to definition are symmetric with respect to the lower indexes

$$K_{kl}^{j} = K_{lk}^{j}.$$

Numbering the phases in the order of fragments sizes decrease

$$d_1 \geq d_2 \geq ... \geq d_N,$$

one should expect

$$K_{kl}^{j} > 0 \text{ for } j > k, l$$
$$K_{kl}^{j} \equiv 0 \text{ for } j < k, l$$
$$K_{kl}^{j} < 0 \text{ for } j = \min(k, l).$$

Finally the mass flux to the j-th phase from the k-th phase being the result of fragmentations could be determined as follows:

$$\kappa_{jk} = \sum_{l=1}^{N_p} S_{kl} v_{kl}^{rel} n_k n_l K_{kl}^{j}(v_{kl}^{rel}, d_k, \rho_k, ...) - \sum_{l=1}^{N_p} S_{jl} v_{jl}^{rel} n_j n_l K_{jl}^{k}(v_{jl}^{rel}, d_j, \rho_j, ...). \quad (3.19)$$

The first term in the right hand side of (3.19) characterizes the positive mass flux to the j-th phase due to collisions of k-th phase particles with the l-th phases. The second term characterizes the negative mass flux due to fragmentation of j-th phase in collisions with other phases.

Subtracting the equation (3.1) multiplied by $\bar{v}_j$ from the equation (3.4) one obtains the momentum equation in the following form:

$$\bar{\rho}_j \frac{\partial \bar{v}_j}{\partial t} + \bar{\rho}_j \bar{v}_j grad\, \bar{v}_j = \vec{F}_j + \vec{F}_{d_j} + \vec{P}_j + \vec{K}_j - I_j \bar{v}_j, \quad (3.20)$$

or

$$\bar{\rho}_j \frac{d\bar{v}_j}{dt} = \vec{F}_j + \vec{F}_{d_j} + \vec{P}_j + \sum_{k=l}^{N_p} \kappa_{jk}(\bar{v}_{jk} - \bar{v}_j), \ \ j = 1, ..., N_p. \quad (3.21)$$

The equation (3.1) makes it possible to determine the number of particles of the j-th phase per volume unit n_j having been transformed as follows

$$\frac{\partial n_j}{\partial t} + div\ n_j \bar{v}_j = \sum_{k=1}^{N} \psi_{jk} + n_{j\ op} + n_{j\ ex} - \eta_j, \tag{3.22}$$

where ψ_{jk} is the number of particles transferred from the k-th to the j-th phase per time unit; $n_{j\ op}$, $n_{j\ ex}$, η_j — the rates of particles number growth and/or decrease due to external sources.

The equations (3.21), (3.22) along with the defined forces and source terms $(\bar{F}_j, \bar{F}_{d_j}, \bar{P}_j, I_j)$ form a closed system of equations enabling to follow the evolution of clouds of fragments taking into account the interactions, breakups, radiation, gravity and aerodynamic drag.

3.3. Marginal Cases Analysis

Sometimes to have a long-term forecast one could simplify the mathematical model. The characteristic period of one cycle for debris clouds and the characteristic times of clouds scattering (about 1 month) are much less than the characteristic time of the global evolution of the debris environment. Then for fragments belonging to the j-th phase, which are characterized by similar size, orbit, inclination, ballistic coefficient etc., their distribution within one altitude layer could be neglected. Then the model provides the possibility to determine the number of particles versus altitude distribution. Averaging the equation (3.22) in longitude and latitude gives the following form of the model equation:

$$\frac{\partial N_j}{\partial t} = -W_j \frac{\partial N_j}{\partial r} - N_j \frac{\partial W_j}{\partial r} + \dot{N}_j, \tag{3.23}$$

where $N_j(t, r)$ is the number of debris particle of the j-th phase per altitude spherical layer of the thickness Δh; $W_j(t, r) = \left(\frac{dr}{dt}\right)_j$— radial velocity of particles (sedimentation velocity); $\dot{N}_j$ — the rate of variation of particles number per altitude layer due to external sources.

$$\dot{N}_j(t,r) = \int_{r}^{r+\Delta h} \int_{-\frac{\pi}{2}}^{\frac{\pi}{2}} \int_{0}^{2\pi} \left(\sum_{k=1}^{N} \psi_{jk}(t,r,\Omega,\theta) + n_{j\ op}(t,r,\Omega,\theta) + \right.$$
$$\left. + n_{j\ ex}(t,r,\Omega,\theta) - \eta_j(t,r,\Omega,\theta) \right) \cdot r^2 \cos\theta\ d\Omega d\theta dr \tag{3.24}$$

The velocity $W_j = \left(\frac{dr}{dt}\right)_j$ is normally negative due to a total tendency to sedimentation of debris particles. Thus some authors prefer to introduce the sedimentation velocity

$V_j(t, r) = -W_j(t, r) > 0$. The sedimentation velocity usually decreases with the increase of an altitude due to a rapid decrease of the drag forces:

$$\frac{\partial V_j(t,r)}{\partial r} < 0. \tag{3.25}$$

Equation similar to (3.23) is the one used for Space debris environment forecasts in (Nazarenko, 1997) and in the Chapter 4 of the present edition.

Another possible simplification is based on the assumption that all the orbits of debris particles are close to circular (Talent, 1990; Nazirov, Ryazanova *et al.*, 1990; Smirnov, Dushin *et al.*, 1993; Ivanov, Menshikov *et al.*, 1996). This approximation relies heavily on tha fact that most of trackable debris objects move along trajectories of very small eccentricities.

Then the momentum equation being of a vector type could be rewritten in two projections: a projection on the radius-vector and a projection on the tangent to the trajectory. The velocity vector for the j-th Lagrangian particle has the following projections in this system of co-ordinates

$$\vec{v}_j = W_j \vec{e}_r + v_{\tau_j} \vec{e}_\varphi = \dot{r}\vec{e}_r + r\dot{\varphi}\vec{e}_\varphi, \tag{3.26}$$

where r, φ — radial and angular coordinates, $\vec{e}_r$, $\vec{e}_\varphi$ — the unit basis vectors; a dot above the symbol denotes a time derivative. The following relationship is valid for debris particles evolution:

$$\frac{|W_j|}{v_{\tau_j}} \ll 1, \tag{3.27}$$

because the sedimentation velocity of debris particles due to atmospheric drag is very low in the upper layers of the atmosphere.

The pressure of sun radiation and the momentum flux due to mass exchange could be also neglected because for LEO the following inequalities are valid:

$$\left|\vec{P}_j\right| \ll \left|\vec{F}_{d_j}\right|; \ \left|v_{jk} - v_j\right| \ll \left|v_j\right| \tag{3.28}$$

The acceleration in (3.21) could be written in the following form:

$$\frac{d\vec{v}_j}{dt} = (\ddot{r} - r\dot{\varphi}^2)\vec{e}_r + (2\dot{r}\dot{\varphi} + r\ddot{\varphi})\vec{e}_\varphi. \tag{3.29}$$

Then the two projections of the vector equation (3.21) look as follows:

$$\bar{\rho}_j(\ddot{r} - r\dot{\varphi}^2) = \bar{\rho}_j\left(\frac{dW_j}{dt} - \frac{v_{\tau_j}^2}{r}\right) = -\bar{\rho}_j g(r) - \frac{1}{2}c_f^j \rho_a(r,t)W_j^2 \frac{3\alpha_j}{2d_j} \tag{3.30}$$

$$\bar{\rho}_j(2\dot{r}\dot{\varphi}+r\ddot{\varphi})=-\frac{1}{2}c_f^j\rho_a(r,t)v_{\tau_j}^2\frac{3\alpha_j}{2d_j}=-\frac{1}{2}c_f^j\rho_a(r,t)(r\dot{\varphi})^2\frac{3\alpha_j}{2d_j} \qquad (3.31)$$

Assuming the orbit evolution caused by the atmosphere to be rather slow, one could neglect the acceleration $\ddot{r}$, as compared with the other terms in the equation (3.30):

$$\ddot{r}\ll r\dot{\varphi}^2.$$

The last term in the right hand part of the equation (3.30) is negligibly small in comparison with the second term in the left hand side due to (3.27). Then the system of equations (3.30), (3.31) takes the form

$$r\dot{\varphi}^2=g(r)\approx\frac{\gamma M}{r^2}, \qquad (3.32)$$

$$2\dot{r}\dot{\varphi}+r\ddot{\varphi}=-\frac{1}{2}c_f^j\rho_a(r\dot{\varphi})^2\frac{3}{2\rho_j d_j}, \qquad (3.33)$$

where γ is the gravity constant and M is the mass of the Earth. Then having determined the angular velocity $\dot{\varphi}$ from the equation (3.32) one could obtain the velocity

$$\dot{r}=W_j=-\frac{3}{2}\frac{c_f^j\rho_a(r,t)}{\rho_j d_j}\sqrt{\gamma Mr}, \qquad (3.34)$$

where the function $\rho_a(r,t)$ could be obtained from one of the models of a standard atmosphere, or from its approximations.

Adopting an approximate formula for density variations with altitude in the upper atmosphere it is possible to simplify more the equation (3.23). Assuming the following formula (Nazirov, Ryazanova et al., 1990):

$$\rho_a(r,t)=\rho_a(r_0,t)\exp\left(-\int_{r_0}^{r}\frac{dr}{H(r,t)}\right) \qquad (3.35)$$

and substituting it in (3.34) one could obtain the following expression for one of the partial derivatives present in (3.23):

$$\frac{\partial W_j}{\partial r}=-\frac{W_j(r,t)}{H(r,t)}\left(1-\frac{H(r,t)}{2r}\right), \qquad (3.36)$$

where $H(r,t)$ is the scale height of a uniform atmosphere at the altitude r; $\rho_a(r_0,t)$ is a known density at the altitude r_0.

Substituting (3.36) in (3.29) brings to the following equation for the particles number density evolution in the atmosphere:

$$\frac{\partial N_j}{\partial t} = -W_j \frac{\partial N_j}{\partial r} + \frac{N_j W_j}{H}\left(1 - \frac{H}{2r}\right) + \dot{N}_j. \tag{3.37}$$

By adjusting the thickness of the spherical layer Δh one could use the equation (3.37) to describe the sedimentation of fragments having elliptic orbits as well. The argument r for elliptic orbits should be regarded as a perigee altitude.

Assuming $\frac{H}{2r} \ll 1$ brings the equation (3.37) to the form used in the Chapter 4.

The last term in the equation (3.37) reflects the particles flux to the j-th phase from the external sources and due to fragmentations in collisions. The external sources should be introduced into the model as additional governing parameters, and the role of those terms would be dependent of the assigned values of the governing parameters, predicting the future human space activities.

Now we'll concentrate our attention on the role of the internal mechanisms of debris self-production in collisions. To understand the role of the term $\sum_{k=1}^{N} \psi_{jk} = \sum_{k=1}^{N} \frac{6\kappa_{jk}}{\pi \rho_j d_j^3}$ in the equation (3.37) one could regard the marginal case $N_p = 1$. Under the assumption there exists only one phase ($\psi_{11} \neq 0$). The particles produced in collisions should be of a smaller size and normally join a different phase. For a one phase model the new-born particles are preserved within the phase thus bringing to a decrease of the mean diameter of fragments.

Then the rate of fragmentation could be introduced by the formula (Smirnov *et al.*, 1993):

$$\psi_{11} = \pi d^2 n^2 v_\tau \Psi, \tag{3.38}$$

where Ψ — the total average number of debris particles generated by one collision to stay in orbit for sufficient time to be taken into account for long-term forecasts.

The initial stages of orbital debris evolution are characterized by rarefied debris environment and the collisions of particles are also rather rare and the average diameter of particles remains stable. The growth of the number of particles due to collisions is proportional to the squared number density (3.38): $\psi_{11} \sim n^2$. After the debris number density increases above the critical value the collisions happen to take place more frequently, the particles number density grows rapidly and their mean diameter decreases. (As it is seen from (3.2) the increase of the number of particles in fragmentation brings to the decrease of the mean diameter characterizing the present phase.) Assuming the new space programs to be stopped by that time one could expect the volumetric content of debris in orbit to be stable (disregarding its gradual decrease due to sedimentation):

$$\alpha = \frac{\pi d^3}{6} n = \text{const (for compact elements)};$$
$$\alpha = \pi \delta^2 dn = \text{const (for plane elements)}. \tag{3.39}$$

Thus the rate of debris population growth turns to be proportional to $n^{2/3}$ for compact elements and $\sim n$ for plane elements, as it follows from (3.38), (3.39). The drag force (3.8) increases with the decrease of particles size thus increasing the rate of sedimentation of small particles contributing to the increase of Low Earth Orbits self-cleaning effect (α decrease).

The equation (3.37) makes it possible to obtain an estimate of the critical number density leading to the cascade effect of debris self-production within a definite altitude layer.

Assuming the number of particles per unit layer to be uniformly distributed one comes to the two competing mechanisms governing the variation of the number of particles within the layer: particles production and self-production on one hand, and self-cleaning due to sedimentation on the other (3.37). then the criterion for the growth of the number of particles will be the following:

$$\frac{NW}{H(r)} + \dot{N}_\Sigma > 0 \tag{3.40}$$

The rate of growth of particles number $\dot{N}_\Sigma$ is the sum of self-production in collisions and production due to external sources (3.24). As production due to external sources is determined mostly by the space policy, this value could be regarded as a slowly varying function with a minor contribution on the eve of the cascade process in comparison with the self-production term. Thus neglecting the external production terms and substituting the expressions (3.34) and (3.38) for sedimentation and fragmentation velocities one could transform the inequality (3.40) as follows:

$$4\pi^2 r^2 d^2 n^2 v_\tau \Psi \Delta h > \frac{3}{2} \frac{N}{H(r)} \frac{C_f \rho_a(r)}{\rho d} \sqrt{\gamma M r},$$

that gives the possibility to evaluate the critical number density (taking into account the relationships: $N = 4\pi r^2 \Delta h n$; $v_\tau = \sqrt{\frac{\gamma M}{r}}$):

$$n > \frac{3}{2\pi d^3} C_f \frac{\rho_a(r)}{\rho} \frac{r}{H(r)} \frac{1}{\Psi}. \tag{3.41}$$

The inequality (3.41) shows that the increase of production per one collision Ψ and particles diameter d brings to the decrease of the critical number density; the increase of the atmospheric density and drag coefficient increases the critical number density. The critical number density decreases with the increase of altitude due to the exponential decrease of density suppressing the other altitude dependent multipliers.

The estimate (3.41) shows that depending on the initial conditions the cascade process of debris growth due to its self-production could start at different altitudes independently. But formula (3.41) is an approximate one based on a number of serious assumptions. Thus it could provide mostly qualitative estimates. To obtain quantitative forecasts one

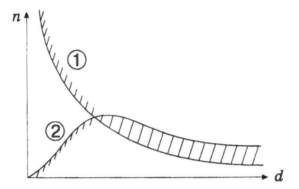

Fig. 3.1. The upper and lower limits for the critical number density of the orbital debris.

needs to integrate the differential equations (3.23) for a number of phases ($j = 1,...,N_p$) accounting for the mass exchange between the phases.

Another qualitative estimate of the long-term scenario of orbital debris evolution based on the inequality (3.41) is the following. The critical number density increases with the decrease of particles diameters ($n_c \geq A/_{d^3}$, curve 1 in Fig. 3.1). The mean diameter of particles decreases in fragmentation bringing to an increase of the particles number density. The last cannot grow faster than

$$n \leq \frac{6\alpha}{\pi d^3},$$ (3.42)

in the absence of new launches.

The volume concentration of debris α decreases due to self-cleaning effect caused by the atmospheric drag. The evolution equation for α looks as follows:

$$\frac{\partial \alpha}{\partial t} = -W \frac{\partial \alpha}{\partial r} + \frac{\alpha W}{H(r)} - \frac{2\alpha W}{r}.$$ (3.43)

Assuming α distributed uniformly one comes to an estimate:

$$\frac{\partial \alpha}{\partial t} \sim \alpha W \left(\frac{1}{H(r)} - \frac{2}{r} \right) < 0$$ (3.44)

always, because $W < 0$ and $H(r)/_r \ll 1$.

Then after a characteristic time t_* passed after termination of space contamination the following estimate for the α decrease is valid

$$\alpha \leq \alpha_0 \exp \left(\frac{W}{H} t_* \right) = \alpha_0 \exp \left(-\frac{3}{2} C_f \frac{\rho_a(r)}{\rho} \frac{r}{H(r)} \frac{v_T t_*}{d} \right) = \alpha_0 \exp \left(-\frac{D_*}{d} \right),$$ (3.45)

where α_0 — the initial debris volume content at termination of external debris production.

The formulas (3.42) and (3.45) allow to obtain the upper estimate for the fragments number density accounting for self-cleaning (curve 2 in Fig. 3.1):

$$n \le \frac{6\alpha_0}{\pi d^3} \exp\left(-\frac{D_*}{d}\right).$$

(3.46)

The estimate (3.46) essentially depends on the value of debris volume content α_0 at termination of external contamination. As it is seen from Fig. 3.1, the decrease of fragments mean diameter being the result of collisions finally makes the actual number density lower than the critical one for the smaller diameters (intersection of curves 1 and 2 in Fig. 3.1). That causes the termination of the cascade process: self-cleaning surpasses self-production. Thus the analysis shows that in the long run self-production should change for self-cleaning of low orbits on termination of space activities. The characteristic times necessary for the self-cleaning process to surpass the self-production could be estimated based on the solution of the unsteady problem using the full set of equations. Nevertheless, the present derivations performed for some marginal cases show that it is never late to undertake the mitigation measures. Mitigation of space contamination in the long run will result in a total self-cleaning of the Low Earth Orbits.

3.4. Some Results of Numerical Investigations

To perform numerical investigations of the developed model (incorporating equations (3.23) and (3.34), for example) one needs to define more accurately the annual rate of origination of new debris particles per altitude layer due to human space activities

$$Q_j = \int_{r}^{r+\Delta h} \int_{-\frac{\pi}{2}}^{\frac{\pi}{2}} \int_{0}^{2\pi} (n_{j\,op}(t,r,\Omega,\theta) + n_{j\,ex}(t,r,\Omega,\theta) - \eta_j(t,r,\Omega,\theta)) r^2 \cos\theta \, d\Omega d\theta dr. \quad (3.47)$$

The formula (3.47) serves, actually, as a forecast of the future space policies based on the analysis of realization of the previous Space programs. An integral evaluation of the parameter Q_j could be given by the following modification of the formula suggested by (Talent, 1990):

$$Q_j = L[(1 - F_e)A_1 D_1 + F_e A_e D_e] \frac{N_j}{N^*}, \quad (3.48)$$

where $N^* = \sum_{j=1}^{N} \int N_j dr$; L — the annual number of launches; A_1 — the average number of operational debris particles originating in one successful launch; A_e — the average number of fragments originating due to one orbital explosion; D_1, D_e — the fraction of fragments remaining in orbit for a long period (more than one year) after a successful launch and after an explosion respectively; F_e — the coefficient characterizing the number of accidental explosions among all the launches. The formula (3.48) being an approximate one, nevertheless, has a sufficient accuracy for the research purposes, but it needs essential further modifications for obtaining quantitative forecasts.

Another aspect to be discussed is that of determining the initial conditions and setting up the initial fragments distribution versus phases. To make our model not so complicated we introduce three phases. The whole number of catalogued objects ($d_1 > 10$ cm) and its distribution versus altitude is introduced by phase one. In addition two phases incorporating untrackable objects are introduced: 1 cm $< d_2 < 10$ cm and 0.1 cm $< d_3 < 1$ cm. The distribution of untrackable objects versus altitude is roughly assumed to be similar to the distribution function for trackable ones.

The collisions of the following type could be taken into consideration: (1,1), (2,2), (1,2), (1,3) and (2,3). The number of particles produced in collisions, was evaluated based on the approximations suggested in (Chobotov, 1990). The contribution from collisions of the particles of the third phase Ψ_{33} could be disregarded as the originated particles were much smaller than those taken into consideration by the model.

To perform numerical investigations the following forecast for the future space activities was adopted: $L = 120$ launches per year; $A_1 = 4$ objects per launch; $A_e = 125$ objects per launch; $D_1 = 0.63$; $D_e = 0.82$; $F_e = 0.03$. The initial distribution of fragments for all the three phases was assumed to be a double pike linear one (see Fig. 3.2 profile at $t = 0$).

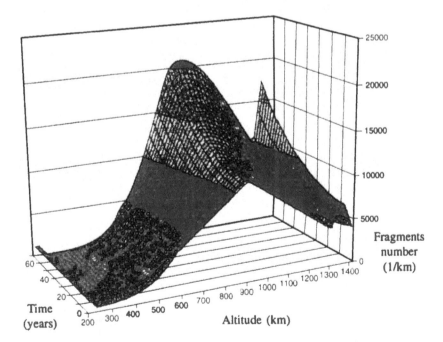

Fig. 3.2. Temporal evolution of fragments number density distribution.

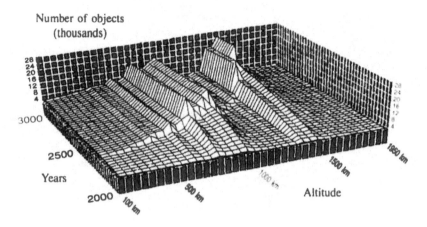

Fig. 3.3. A version of a long-term forecast of the orbital debris distribution.

The calculations were performed within the altitude range: 240 km $< h <$ 1400 km. The Fig. 3.2 shows the temporal evolution of the initial spatial distribution of fragments belonging to the phase 3 versus altitude (Lim, 1995). The larger fragments belonging to phases 1 and 2 practically did not change their spatial distribution within the relatively small time interval regarded. The only changes that could be distinguished for the phase 2 took place within the interval $h < 600$ km, and for the phase 1 – $h < 450$ km.

The phase 3 spatial distribution underwent essential changes within the 70 years time interval regarded. At lower altitudes $h < 550$ km an essential decrease of the debris population took place. Periodical oscillations of the fragments number at those altitudes are due to the 11-years cycle of solar activity bringing to periodical increase of the density ρ_a at those altitudes. For higher altitudes (600 km $< h <$ 800 km) the amplitudes of the oscillations decrease and the mean decrease of the number of fragments taking place within the first years changes for the growth of fragments number due to collisions and fragmentation of the larger particles belonging to other phases. For higher altitudes the growth of fragments number takes place.

The temporal evolution of the spatial distribution for the number of fragments belonging to phases 1 and 2 versus altitude (Smirnov et al., 1993; Ivanov, Men'shikov et al., 1996) with an interval 50 km is shown in Fig. 3.3. The figure shows that the increase of the number of large fragments within the first two hundreds of years is rather slow. Nowadays the number of debris fragments in orbits is far from being critical. After the critical conditions are reached the number of collisions grows rapidly and the debris self-production process becomes dominant. It could be seen that for different altitudes the critical conditions are reached at different times. The process of debris self-production begins later at higher altitudes that is caused by peculiarities of initial fragments distribution.

After the critical concentration of fragments per volume unit n_c is reached the orbit could no longer be used for space flights. Since then the new launches on the orbit are supposed to be stopped and the external contamination of the orbit is assumed to be terminated. Fig. 3.3 shows that after termination of external contamination the rapid growth of the debris population changes for its slow reduction due to self-cleaning process. Nevertheless, the reduction of debris population is a very slow process, as compared with its increase due to self-production. Thus the orbit remains hazardous for space flights for a very long period.

The results illustrated by the Figs. 3.2 and 3.3 should be treated as a qualitative forecast and serve just to illustrate the properties of the developed model. To obtain a quantitative forecast a more detailed models for debris production due to fragmentations in collisions and orbital explosions should be incorporated into the model. Those models will be discussed in the Chapter 7 of the present book.

3.5. Conclusions

The developed mathematical model for space debris evolution is based on a continua approach thus serving an alternative to the classical celestial mechanics approach the existing debris evolution models are based on. The model is able to follow the evolution of multicomponent debris clouds incorporating classes of fragments of different types.

The analysis of the marginal cases made it possible to derive a simplified criterion determining the critical number density characterizing the beginning of the cascade process of debris self-production in collisions. It was shown that, contrary to a predominant point of view, collisions of different types of fragments contribute not only to debris self-production, but to a self-cleaning of the low Earth orbits, because fragments of small diameters are much stronger influenced by the atmospheric drag.

Numerical investigations of the model showed its sensitivity to variations of the external governing parameters. It was demonstrated that the cascade process of debris self-production could start at different altitudes non-simultaneously. To obtain quantitative forecasts for orbital debris long-term evolution it is necessary to incorporate more detailed models describing the breakups variety into the developed model of debris evolution.

CHAPTER 4

THE SOLUTION OF APPLIED PROBLEMS USING THE SPACE DEBRIS PREDICTION AND ANALYSIS MODEL*

A.I. Nazarenko

The present chapter illustrates the applications of space debris prediction and analysis models to solving applied problems of collisions of spacecraft with debris particles assessment accounting for shape and orientation of typical spacecraft modules. The model takes into account the spatial density of distribution of particles larger than 0.1 cm and its evolution. Examples of long-term forecasts of orbital debris environment evolution are given for different scenarios of future human space activities.

4.1. Basic Principles of the Model Construction. The Data on New Yearly Formed Objects

4.1.1. Introduction

The technogeneous contamination of the near-Earth space (NES) is characterized by very limited possibilities of its state monitoring. Reliable data are available only for large (catalogued) objects sizing 10–20 cm. The data on smaller fragments were obtained only in small regions of a multidimensional space: the object size — altitudes — time. This is clearly seen from Fig. 4.1 (UNCOPUOS, 1998). By this reason the construction of

---

* The present research was supported by the Russian Foundation for Basic Research (project 98-01-00218).

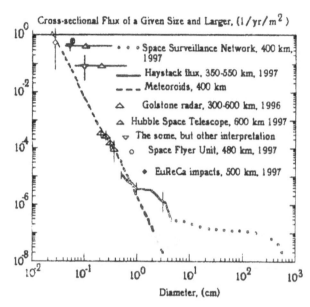

Fig. 4.1. Flux of debris and meteoroids.

reliable estimates of a current state of technogeneous contamination and its forecasting are extremely difficult problems.

Obviously, if detailed data about orbit elements of small objects are absent, it is necessary to apply statistical approach for studying danger of spacecraft collisions with them. At the same time traditional approach to studying satellites motion is deterministic. It is based on integrating motion equations. In that way working out new models for the motion of large groups of satellites, on the one hand, one should take into account their motion regularities and, on the other hand, the model must be based on statistical description of initial conditions.

In the majority of papers ideas of deterministic approach are used. For example, in modern NASA model (Reynolds, 1991) after-effects of all known launching and destruction of satellites and possible future analogous events as well are imitated. For each object (or a group) a vector of initial conditions is formed. Forecasting is performed using the traditional motion models. To estimate the collision danger of a pair of satellites the methods of D. Kessler are used (Kessler, 1981). The results are summarized for a great number of objects. Obviously, this approach is very labor-consuming: it can be realized only by using powerful enough computers. In spite of its labor-consuming character, this approach doesn't solve a problem of guaranteeing the model's adequacy. Modeling accuracy of destruction after-effects is unknown. In the article (Bade, 1996) it is noted that errors can be 10-multiplying. The model tuning according to available measuring data turns to be a rather difficult task.

There are (Talent, 1990; Ivanov, 1996) rather simple analytical methods of statistical description of space debris (SD), but simplifications introduced into the models do not permit to get an adequate estimate of the current situation: their results are not applied in the most well-known models (Jehn, 1997).

The only possible approach to solving this problem consists in modeling the techno-geneous contamination based on using all a priori and experimental data. The important role of mathematical modeling in analysis of the situation in the near Earth space environment is defined by the following circumstances:

- In view of impossible cataloguing of small fragments (less than 10 to 20 cm in size) their statistical spatial distribution model appears to be virtually the only effective basis for solution of a broad range of applied problems;
- It is only the mathematical model that could estimate probable grave collision of a given spacecraft (SC) with debris fragments with a sufficient accuracy;
- Only an adequate mathematical model could produce the future NES contamination assessment — for decades and hundreds of years in advance;
- The development and tests of various measurement facilities and detectors are unimaginable should the corresponding mathematical models be ignored;
- The correlation of (rather limited) experimental data with modeling results is the efficient and, virtually, unique method of obtaining information about the evolution of contamination process in the regions, where the experimental data are not available.

The conventional approach to prediction of technogeneous space contamination is based on modeling SD generation as a result of satellite launches, technological operations and spacecrafts breakups. The detailed data about all known events are used on a preceding time interval. The scenarios of possible operations and fragmentation are simulated for future time moments.

4.1.2. Approach

The basic principles of applied here statistical description of the motion of satellites ensemble are (Nazarenko, 1997) the following:

- technogeneous environment is characterized by spatial distributions of objects spatial density and values and directions of their velocities;
- regularities of objects motion as the Earth's satellites are applied;
- a priori data about perturbing factors is made use of as much as possible;
- an averaged description of pollution sources is performed allowing to minimize the number of parameters, that are defined more exactly on the known experimental data.

The main cause for small- and medium-size SD formation is orbital fragmentation of space objects. We apply the averaged approach to the description of SD sources. Its characteristic feature is that we apply, instead of the data on specific launches and break up cases, the following averaged data: (a) the altitude distribution $dp(t, h_p, d)$ of a number of annually formed objects sizing larger than d (here t is time, h_p is the perigee altitude), and (b) statistical distributions of their eccentricities $p(e, d)$ and inclinations $p(i, d)$. The justification of such an approach is as follows:

– The SD number varies insignificantly during a year (a few percents only). Therefore, the more detailed (in time) modeling of SD sources is excessive: it strongly complicates the model, practically not influencing the accuracy.
The note. This statement does not exclude the possibility and expediency of detailed modeling of break up consequences at short time intervals, when the SD "cloud" remains rather compact. However, the cloud scattering process is known to proceed rather quickly, as a rule. The duration of this process is about 1 month.
– The reasons and conditions of satellites fragmentations, which resulted in formation of the majority of small SD fragments, are extremely diverse. Therefore, it is difficult to expect that the results of modeling consequences of all known fragmentation (the number of particles, the fly-away velocity) are accurate enough. The level of errors of such a modeling is unknown. Hence, the approach, based on averaged data, does not seem to be worse, but looks even more preferable.
– The previous statement is even more valid for the future moments of time, for which the reasons and circumstances of fragmentations are unknown. The use of the averaged data on the intensity of new SD formation seems to be optimal for performing the forecasts.
– One more reason in favor of applying the averaged data on new SD formation intensity is based on the fact, that the averaged data contain a smaller number of parameters in comparison with the detailed data. It is known that the minimal number of any forecasting model parameters to be updated on the experimental data basis is desirable under the conditions of limited measurement information. Therefore the model having a smaller number of parameters to be tuned will provide better forecasting accuracy in comparison with models having a larger number of such parameters.
– Setting the distribution of three orbit elements (h_p, e, i) is quite enough for statistical description of the situation because the other tree elements (the longitude of the ascending node Ω, the perigee argument ω and the average anomaly at the initial moment M_0) sufficient for enclosures precision are distributed uniformly on the interval $(0–2\pi)$.

The dependences of initial distributions on objects' sizes are constructed based on the natural assumption, that all small-size SD were formed as a result of large (catalogued) objects fragmentation. Such an approach is based on using: (a) Monte-Carlo method, (b) statistical distributions of annual surplus $dp(h_p, d_{cat})$, $p(e, d_{cat})$, $p(i, d_{cat})$ constructed from the real data on catalogued objects, and (c) a priori data about the dependence of SD fly-away velocity on their size. The less is the size of a particle, the greater velocity increment it receives at the formation time. In such a way the initial distributions are updated by using the fragmentation modeling data.

In a number of works (Talent, 1990; Ivanov, 1996) an assumption is used, that all orbits are similar to circular. Let us examine possible after-effects of this simplification. In Fig. 4.2 eccentricities distributions of the SO of various sizes are shown. The results shown in Fig. 4.2 were obtained under the assumption that fragmentation events having caused the formation of those small-sized objects has taken place in circular orbits. Nevertheless,

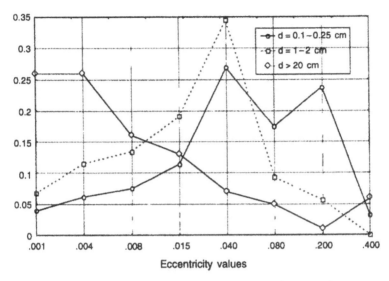

Fig. 4.2. Distribution of eccentricities for SD of various size.

one can see that a share of objects with larger eccentricities increases for smaller fragments in comparison with the corresponding distribution of catalogued SO.

In Fig. 4.3 the estimates of SO spatial density at various altitudes are shown formed by objects, which perigee altitude is placed in one of altitude ranges (under 800 km, from 800 to 1300 km and more than 1300 km). The data show that the influence of objects with the perigee placed in a definite range goes far beyond the range limits and is spreading

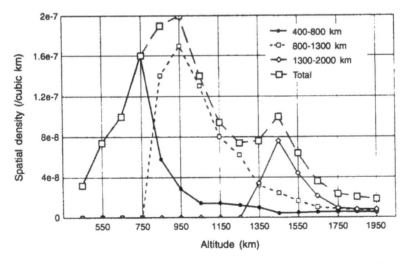

Fig. 4.3. The contribution of various SD groups (sizing 1.0–2.0 cm) into the total altitude dependence of the spatial density.

at higher altitudes. For example, at the altitude of 850 km the spatial density of objects with perigee altitude below 800 km is 35% of their spatial density at the altitude range of 700–800 km. The considerable part of non-circular orbits is the cause of this effect. About a half of all objects have orbits eccentricities more than 0,01. In that way an assumption about the circular form of all orbits could lead to significant errors.

4.1.3. Principles of the Space Debris Environment Forecast

Let us consider various space objects (SO), characterized by orbit perigee altitudes not exceeding 2000 km, and choose the perigee altitude (h_p) from the vector of SO orbital elements. We shall suppose that among all variable SO parameters only the perigee altitude essentially influences the evolution of altitude distribution of SO number. The other orbital elements will be designated by χ. We sub-divide the whole set of objects with different elements χ into some finite number of sub-sets (groups) with elements $\chi_1, l = 1, 2, \ldots l_{max}$. Let $p(t, h)$ be the density of the perigee altitude distribution for objects from the selected group at time t. Then we state the problem of studying the laws of density variation in time. Index l will be omitted hereafter in analyzing the distribution evolution for some particular SO group.

The partial derivative equations are derived for describing the evolution of altitude distribution of SO number:

$$\frac{\partial p(t,h)}{\partial t} = V(t,h)\left[\frac{\partial p(t,h)}{\partial h} - \frac{p(t,h)}{H(t,h)}\right] + dp(t,h,\ldots). \tag{4.1}$$

Here $V(t,h)$ is the perigee lowering velocity; $H(t,h)$ is the atmosphere scale height; $dp(t,h,\ldots)$ is the rate of SD density increment due to various reasons.

In calculating the evolution of altitude distribution SO number the following factors are taken into account:

- the atmospheric drag at altitudes of up to 2000 km;
- the sub-division of all SO by parameters into the groups which differ in size d, eccentricity e and ballistic factor S;
- the initial altitude distribution of SO of various types;
- the expected intensity of formation of new SO of various types as a result of launches and explosions: $dp(t,h,\ldots)$ is the increment of SO number at various altitudes per time unit;
- the non-stationary character of factors taken into account, namely, the atmospheric density in connection with solar activity variation during the 11-year cycle and the intensity of new launches.

The perigee altitudes are of special importance among the listed orbital parameters taken into account. While rendering the basic influence on orbit lowering, it changes (the perigee altitude is lowered finally bringing to SO re-entry). Therefore, the perigee altitude

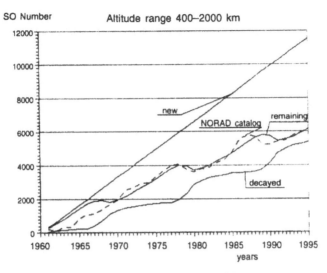

Fig. 4.4. Forecasts of the catalogued SO number.

is one of arguments of equations (4.1). The ballistic factors and SO size have absolutely different character of influence: they do not practically change during the orbital evolution process. The eccentricity has the intermediate character of influence: it changes, generally speaking, under the atmospheric drag action, but this change does not play essential role, because the large part of SO have orbits with low eccentricities.

The algorithm for numerical solution of equations (4.1) is developed (Nazarenko, 1993). The compromise between the detailed investigations and the simplicity of an algorithm is required in choosing a number of SO subdivision into groups. If the sub-division is too detailed, there will not be enough memory and the calculation time will increase. Besides, one should take into account, that the initial data for the environment forecasts have a rather high uncertainty, that makes the detailed algorithm senseless.

4.1.4. The Data on New Yearly Formed Objects

The most reliable data exist for generation of large catalogued objects. These data present objects distribution over perigee altitudes, eccentricities, inclinations and other parameters. The yearly average number of launches on a preceding interval was about 100. The average number of catalogued objects per 1 launch was 7.5, and 2.5 of them were technological elements, 5 were break up fragments. The yearly average number of catalogued objects per 1 break up was 120.

These data were used as a basis for constructing the right-hand side of equation (4.1) for catalogued objects. Fig. 4.4 presents corresponding results of forecasting the situation on a preceding time interval. The data of this figure show that a rather good agreement between the model and experimental NORAD data is provided here.

Table 4.1. The ratio of a number of SOs larger than d_j in size to catalogued SOs number ($j = 8$)

j	1	2	3	4	5	6	7	8
d_j, cm	0.10	0.25	0.50	1.0	2.0	4.0	8.0	20
Relation	13 500	1 500	316	54	21	8	3.5	1.0

A much more difficult problem, however, is the construction of right-hand sides of equation (4.1) for small-sized, non-catalogued objects. The distinguishing feature of the applied approach for description of pollution sources is in using the following averaged data instead of the data on specific launches and break up incidents:

- An averaged number of annually generated objects of various sizes $n(d > d_j)$.
- The ratio $k(d_j)$ of a number of annually generated small-size objects and corresponding number of catalogued SO

$$n(d > d_j) = k(d_j) \cdot n(d_{cat}). \qquad (4.2)$$

These data are given in Table 4.1.
- The coefficient of technical policy K, that characterizes the ratio of annually generated number of catalogued SO at future time moments to corresponding estimation at the preceding interval.
- The altitude distribution of a number of annually generated objects $dp(h_p, d)$ sizing larger than d (here h_p is the perigee altitude), and statistical distributions of their eccentricities $p(e, d)$ and inclinations $p(i, d)$.

The estimates presented in the Table 4.1 are adjustable parameters of the model. They are updated in such a manner, that the acceptable correspondence between modeling results and experimental data should be provided (Nazarenko, 1995). In so doing we used the data of Fig. 4.1, as well as the published materials of Haystack (Stansbery, 1996) and FGAN (Leushake, 1997) radar measurements, etc.

The analysis has shown that the accuracy of modeling results is not worse than the accuracy of known analogies (Jehn, 1997). Further on, the model parameters will be updated as the new experimental data are received.

4.1.5. Distribution of Parameters of New-Formed Objects

According to available data, the generation of small-sized non-catalogued objects is mainly the result of accidents (breakups, explosions) in orbits. The smaller is the particle, the greater is the velocity increment it acquires on being generated. The distributions $p(h_p)_j$ and $p(e)_j$ of non-catalogued SD of various size should be updated using the fragmentation modeling data.

We shall make use of available fragmentation models by applying the obtained dependences of maximal velocities of particles fly-away on their size $V_{max}(d_j)$ and the probabilistic law of distribution of a fly-away velocity $p(V/V_{max})$. These functions are regarded as initial ones and could be varied in the course of investigations.

Here the assumption is used, that all breakups take place on circular orbits only. As a result of velocity changing, the orbits of newly generated objects become elliptical. The change of perigee (or apogee) altitude depends on decreasing (or increasing) of the tangential velocity component V_τ:

$$\Delta h \approx 4a\frac{\Delta V_\tau}{V}, \quad e \approx 2\left|\frac{\Delta V_\tau}{V}\right|. \qquad (4.3)$$

Here a and V are the values of a semimajor axis and velocity for an initial orbit.

The calculations of increments of elements h_p and e for orbits with various perigee altitudes as well as for various random values and directions of particles' fly-away velocities, allow to construct the laws of distribution of these increments for various conditions: $p(\Delta h, h_p)_j$ and $p(e, h_p)_j$. Once these functions are constructed, the laws of distribution of elements under consideration after the fragmentation could be calculated using the following formulas:

$$p(e)_j = \int_h p(e,h)_j \cdot p(h)_c \cdot dh,$$

$$p(h_p)_j = 0.5 \cdot \int_{h>h_p} p(h-h_p,h)_j \cdot p(h)_c\, dh + 0.5 \cdot p(h)_c. \qquad (4.4)$$

Here function $p(h)_c$ is the distribution of perigee altitude before a fragmentation event (for catalogued objects). The integrals (4.4) are calculated in the form of discrete functions, using the histograms appearing in the right-hand sides of equations.

We constructed (Nazarenko, 1997) the distributions of perigee altitudes and eccentricities of fragments at the time of their generation (Fig. 4.5 and Fig. 4.2, correspondingly). The range of possible values of eccentricities in Fig. 4.2 is subdivided into unequal intervals. It can be seen from the figure that the less is the size of particles, the greater is the portion of objects with high eccentricities.

The modeling results showed that for fragments larger than 1 mm in size at fly-away velocities less than 4 km/s the distributions of fragments' inclinations only slightly differed from corresponding distributions of catalogued objects.

4.1.6. General Characteristics of SD Software

Using the approach presented above, as well as taking into account some other investigations, the author developed the computer model for SD Prediction and Analysis (SDPA-model). General data about this Model: this is a semi-analytical stochastic model for

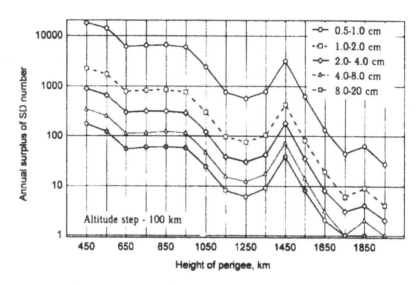

Fig. 4.5. Perigee height distribution of the annual surplus.

Table 4.2. *Principal characteristics of SDPA Model*

No	Debris modeling characteristics	Comments
1	Range of altitudes	up to 2000 km.
2	Range of sizes	more than 1 mm.
3	Forecast of the environment	Fulfilled
4	Description of debris sources	Averaged
5	Construction of a spatial density distribution	Fulfilled
6	Construction of a statistical velocity distribution	Fulfilled
7	Estimation of possible collisions' characteristics	Fulfilled for the given SC
8	Account of spacecraft shape and orientation	Fulfilled
9	Users' possibilities for solving the above mentioned problems items 3, 5–8	Fulfilled
10	Computer implementation	PC
11	Solution time	Less than 1–2 minutes
12	Possibilities of using	Software

medium and long term forecast of the LEO debris environment, for construction of spatial density and velocity distributions as well as for the risk evaluation (Table 4.2).

The example of one of sections of the model menu is shown in Fig. 4.6.

This model is implemented on a personal computer and was used for investigations by a great number of firms and enterprises in Russian Federation.

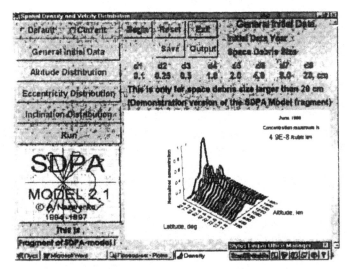

Fig. 4.6. Example of SDPA model menu.

4.2. Current Space Debris Environment Containing Particles Larger Than 0.1 cm

4.2.1. Introduction

In accordance with statistical approach to the description of space debris environment, it is characterized by: a) the dependence of a spatial density on the altitude and latitude of a point, as well as on the size of objects $\rho(h, \varphi, d)$ and b) by statistical distributions of the magnitude ($p(V,...)$) and direction of space object (SO) velocity ($p(A,...)$) in the inertial geocentric coordinate system. The longitudinal dependence is not taken into account, as a rule, since the statistical distribution of the ascending node longitude of all SOs in the LEO region is close to uniform one.

Two approaches are usually applied to SO spatial density distribution construction. The simplest one is the numerical solution based on the total SO catalogue information: all SOs are forecasted for some moment of time, and the number of objects falling into specified areas of space is counted. The averaging of these results over a set of time moments allows to construct the statistic spatial distribution. This techniques yields rather reliable and realistic results, but has some drawbacks as well: it is not adjusted to analysing the future hypothetical situations and small non-catalogued fragments' distribution.

For application of the traditional approach the 6-dimensional vector of initial conditions of all objects is to be known. This is not always possible due to many reasons. First, not all researchers have access to the full catalog of elements of all objects sizing larger than 10–20 cm, and, second, for objects of lower size the real elements of orbit are not known at all.

In many cases, the initial data are known not as particular initial conditions, but rather in the form of statistical distributions of basic parameters, such as the perigee height, eccentricity, inclination, etc. By this reason the development of the techniques for construction of SO concentrations is a topical problem.

The analytical techniques for the construction of SO spatial density distribution, which is free from the above-mentioned drawbacks of numerical approaches, is outlined in papers by D. Kessler (Kessler, 1981) and A. Nazarenko (Nazarenko, 1993). The first paper's approach is based on the fact, that in calculating the probability of collisions of SOs pair the spatial density is constructed for each of the objects. The second paper considers the construction of density $p(h, \varphi)$ as an independent problem, that is solved analytically for the whole set of SOs. In so doing the initial SO data are specified in the form of statistic distributions of orbital elements. Therefore, the latter techniques can be considered to be some generalisation of D. Kessler's approach. In the remaining aspects the differences between these two approaches are not principal.

4.2.2. Spatial Density Distribution Construction Technique

This section outlines the technique for construction of SO spatial density $p(h, \varphi)$ as a function of latitude φ and altitude h. The initial data required for spatial density construction are also presented.

Let us consider in more details the basic principles of this technique being the most suitable for analysing the non-detectable space debris populations. We shall use only two arguments of spatial density: altitude h and geographic latitude φ. The independence of density on the longitude follows from a virtually uniform distribution of SO nodes' longitude within the altitude range under consideration. We introduce into consideration the statistic density of SO distribution over altitude and latitude:

$$p(h,\varphi) = \frac{\partial^2 N(h,\varphi)}{\partial h \partial \varphi}. \qquad (4.5)$$

Here $N(h, \varphi)$ is the number of SOs with current values of altitude h' and latitude φ', which are inside the region: $h' < h$, $\varphi' < \varphi$. Under the assumption, that the longitudinal distribution of SOs is uniform, the knowledge of distribution $p(h, \varphi)$ allows easily to determine the spatial density of SOs number:

$$\rho(h,\varphi) = \frac{p(h,\varphi)}{2\pi \cdot (h+R)^2 \cdot \cos\varphi}. \qquad (4.6)$$

Here R is the Earth radius.

In constructing the spatial distribution the following data are used as the initial information:

1. N_Σ — the total amount of SOs with perigee height h_p lower than some specified value h_{max}.

2. $p(h_p)$, $p(e)$, $p(i)$ — the densities of distribution of values of perigee height, eccentricity (e) and inclination (i) of space objects in the region under consideration. For the above-mentioned densities, which are supposed to be independent, the following relations are valid:

$$\int_{h_p} p(h_p) \cdot dh_p = 1; \quad \int_e p(e) \cdot de = 1; \quad \int_i p(i) \cdot di = 1. \tag{4.7}$$

The function $p(h, \varphi)$ is constructed in two stages. Let us consider the spherical layer with altitude boundaries $(h, h + \Delta h)$. First, we determine the average number of objects $\Delta N(h, h + \Delta h)$, which are inside the mentioned spherical layer at some fixed moment of time. At the second stage we construct the distribution $\Delta N(h, h + \Delta h, \varphi)$ of objects in the given layer over the latitude. This distribution satisfies the relationship

$$\int_\varphi \Delta N(h, h + \Delta h, \varphi) \cdot d\varphi = \Delta N(h, h + \Delta h). \tag{4.8}$$

Taking it into account, we obtain

$$p(h, \varphi) \cdot \Delta h = N_\Sigma \cdot \frac{\cos\varphi}{\pi} \cdot F(\varphi) \cdot \int\int_{h_p\,e} \Delta\tau(h_p, e, h) \cdot \Phi(h_p, e, h) \cdot p(h_p) \cdot p(e) \cdot dh_p \cdot de. \tag{4.9}$$

Here: $F(\varphi) = \int_i \dfrac{p(i) \cdot di}{\sqrt{\sin^2 i - \sin^2 \varphi}}$, for $\sin i \geq \sin \varphi$. $\tag{4.10}$

Quantity $\Delta\tau(h_p, e, h)$ has a meaning of probability of finding SOs with orbital elements h_p, e, h in the spherical layer under consideration. It is determined using the technique presented in the paper (Nazarenko, 1993). The function $\Phi(h_p, e, h)$ is the following:

$$\Phi(h_p, e, h) = \frac{(1-e)^2}{\sqrt{1-e^2}} \cdot \left(\frac{h+R}{h_p+R}\right)^2. \tag{4.11}$$

In the relation (4.9) the integrals with respect to h_p and e arguments are taken throughout the region of their possible values. In accordance with dependence (4.6), based on the function $p(h, \varphi)$ one can easily calculate the unknown function $\rho(h, \varphi)$, that characterises the altitude and latitude dependence of a number of SOs per unite of volume.

In accordance with the above technique, the author has developed a program for construction of SO spatial density as a function of altitude and latitude in the near-Earth space. The $N_\Sigma \cdot p(h_p)$, $p(e)$, $p(i)$ distributions are used as initial data, which are implemented as histograms and immersed into the initial data files. The program allows to construct a spatial distribution for any SOs: both catalogued and non-catalogued ones. All initial

Table 4.3. Sub-division of catalogued SOs into groups

	Group No					
	—	1	2	3	4	–
Altitude, km	<175	175–800	800–1300	1300–2000	175–2000	>2000	Total number
Eccentricity, e	—	< 0.5	< 0.5	< 0.5	>=0.5	—	of SOs
Number of SOs	28	2212	2591	1279	664	1149	7867

data written down into the above-mentioned files is managed by the user: the initial data can be easily corrected or fully substituted by the user either manually or by means of individual user's programs.

Below the results of analysis of characteristics of these SOs are presented taken from the catalogue data for June, 1996.

To take into account the dependences of $p(i)$ and $p(e)$ distributions on height h_p, we analyzed these distributions in details at various altitudes and found the altitude ranges, inside which these distributions may be considered independent of the perigee height. The solution of this problem is based on applying a specially developed program of constructing histograms of i and e elements from the data of a current SO catalogue.

In accordance with the values of altitude and eccentricities, all SOs were subdivided into four groups. The data on sub-division and a number of objects are given in Table 4.3.

The number of objects in four groups under consideration was 6746. This equals 85% of the total number of SOs in a catalogue.

Fig. 4.7 shows the distribution of SO's perigee heights in the altitude range of 400 to 2000 km, and Figs. 4.8 and 4.9 give the distributions of inclinations and eccentricities for SOs from various groups.

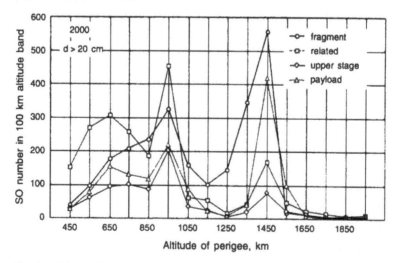

Fig. 4.7. Distribution of perigee altitudes for various types of catalogued SOs.

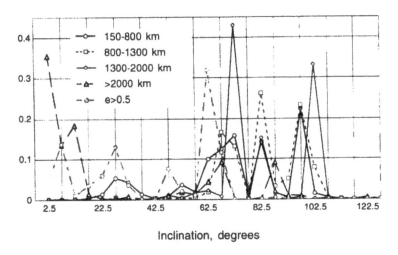

Fig. 4.8. Inclination distribution of SO of various altitude groups.

The distributions of inclinations differ most essentially between various groups. The maximum is prominent in the regions of inclinations of 72.5 and 102.5 degrees for SOs from Group No. 3. For Group No. 2 the distribution density maximum is achieved in the other regions of inclinations: of 82.5 and 97.5 degrees. The distribution of inclinations for SOs from Group No. 1 is more "blured", and in Group No. 4 two maxima are prominent: at inclinations of 2.5 and 62.5 degrees.

It is seen from Fig. 4.9, that the distributions of eccentricities for SOs from Groups Nos. 1, 2, and 3 do not essentially differ from each other. The only exception are the objects with perigee heights greater than 2000 km, for which the eccentricities lower than 0.004 are prevailing.

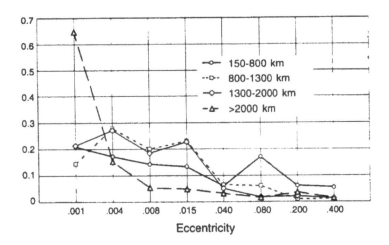

Fig. 4.9. Distribution of eccentricities of SO of various altitude groups.

Probably, the above distributions do not take into account some features of a non-catalogued SOs generation. But at present time these assumptions are quite reasonable. Their application is also justified due to the fact, that a considerable fraction of catalogued objects represent those generated by explosions and fragmentation, i.e. under the same conditions as the major part of non-catalogued SOs.

The basic problem of the initial data preparation for constructing a spatial distribution of non-catalogued SOs is the distribution of perigee heights $N_\Sigma \cdot p(h)$. These distributions drastically differ for catalogued and non-catalogued SOs. The differences relate both to the number of objects N_Σ and to the form of distribution $p(h)$. The latter is due to essential distinctions in ballistic factors for two SO types under consideration. These distinctions result in different rates of NES "self-purification" process: this process is much more faster for non-catalogued SOs. All these issues are outlined in more details below.

4.2.3. Velocity Distribution Construction Technique

This Section outlines the technique for constructing statistic distributions of SO velocity vector magnitude and direction as functions of altitude and latitude in the near-Earth space. The flux of particles has a meaning of the product of their density (the number of particles per volume unit) by the relative velocity. The main difficulty in constructing the model of space debris fragments' flux is due to the fact, that at any NES point the magnitude and direction of velocity vectors are not fixed (constant). Their possible values lie within a rather wide range. Thus, the problem is reduced to constructing statistic distributions of SO velocity vectors magnitude and direction as functions of altitude and latitude in the NES.

Let us first consider the distributions of tangential and radial velocity components. The technique of their construction is based on the algorithms described in the previous Section. The technique is the following. For any spherical layer $(h, h + \Delta h)$ the values of orbital elements h_p and e from the region $(h_p, h_p + \Delta h_p)$, $(e, e + \Delta e)$ correspond to the fixed values of tangential and radial velocity components $V_\tau(h_p, e)$ and $V_r(h_p, e)$, as well as to the certain probability of hitting into the mentioned region

$$P(h, h_p, e) = \Delta\tau(h_p, e) \cdot p(h_p) \cdot p(e) \cdot \Delta h_p \cdot \Delta e .$$ (4.12)

This allows us easily to construct the histograms of velocity components' distributions and calculate statistic characteristics of these distributions: the mean value, variance, etc.

Some results of constructing the distributions of radial and tangential velocity components at various altitudes are given in Figs. 4.10 and 4.11. It is seen that the radial velocity component does not exceed 0.1 km/s in 90% of cases. This corresponds to ±0.8 deg. deflection of velocity vector from the horizontal plane. The tangential velocity component greatly depends on the altitude (it decreases as the altitude grows), but at some fixed altitude it varies within rather narrow limits: in the range of 0.25 km/s in 90% of all possible cases at least.

Now we shall pass to studying possible directions of a tangential component of the velocity vector.

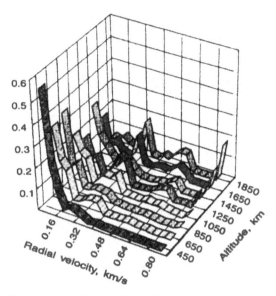

Fig. 4.10. Distribution of a radial velocity component.

Let us consider the arbitrary point of the near-Earth space with spherical coordinates r, φ, λ. Assuming the SO spatial density $\rho(r, \varphi)$ to be known, we determine the number of objects, which pass in this point's vicinity through the cross-section of size δF per the time of one period (one revolution), and construct the azimuthal distribution of frequency (probability) of such passages $p(A)$.

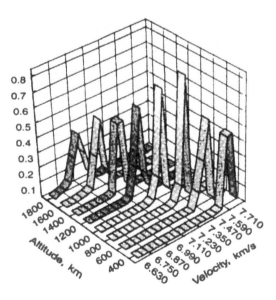

Fig. 4.11. Distribution of a tangential velocity component.

– The total number of objects being in the altitude layer $(r, r + \delta r)$ is $\delta n = p(h) \cdot \delta r$, where

$$p(h) = 2\pi \cdot r^2 \cdot \int_{-\pi/2}^{\pi/2} \rho(h, \varphi) \cdot \cos \varphi \cdot d\varphi. \qquad (4.13)$$

Only a small fraction of δn passes in the given points' vicinity. The problems is: to determine the number of objects which pass the assigned point at a distance δb not exceeding:

$$|\delta b| \leq \delta r / 2. \qquad (4.14)$$

This condition (4.14) being satisfied depends on two elements: inclination i and ascending node longitude Ω (that will be counted clockwise from the given point's longitude). For the trajectories passing strictly through the given point these elements satisfy the relation

$$\mathrm{tg}\, i \cdot \mathrm{tg}\, \Omega = \mathrm{tg}\, \varphi. \qquad (4.15)$$

If for any other value of Ω and $i = f(\Omega)$ we determine such deviations $\delta\Omega$ and $\delta i = F(\delta\Omega)$, that the condition (4.14) is valid, then it will be possible to determine quantitatively the sought fraction of objects from the a priory specified distribution $p(i)$ and $p(\Omega)$ and from the relationship:

$$\delta n(\delta\Omega) = p(i) \cdot p(\Omega) \cdot \delta i \cdot \delta\Omega. \qquad (4.16)$$

This is just the fraction of objects from the δn number, which have the node longitude within the $(\Omega, \Omega + \delta\Omega)$ interval and pass through the δb-vicinity of the given point. There exists a single-valued correspondence between the elements and velocity vector direction azimuth, namely

$$\sin A = \frac{\cos i}{\cos \varphi} = \frac{\sin \Omega}{\sqrt{\sin^2 \Omega \cdot \cos^2 \varphi + \sin^2 \varphi}} \qquad (4.17)$$

The value of azimuth relates to the same quadrant, as the ascending node longitude value.

The basic problem consists in finding, for the set of trajectories $i = f(\varphi, \Omega)$, the values δi and $\delta\Omega$ (the "tube" of trajectories), for the condition (4.14) to be satisfied. Once this region S is constructed, the curvilinear integral

$$\delta n(\delta b) = \int_S p(i) \cdot p(\Omega) \cdot dS \qquad (4.18)$$

will determine the fraction of objects from the δn number, which are situated inside the "tube" mentioned above. As a result of this "tube" construction, the curvilinear integral could be presented as an ordinary integral

$$\delta n(\delta b) = \frac{\delta b}{r \cdot \sin \varphi} \cdot \int_0^{2\pi} p(\Omega) \cdot p(i) \cdot \sin i \cdot d\Omega. \tag{4.19}$$

The inclination i in the integral is associated with the ascending node longitude by the relation (4.15). The total number of objects, which pass in the given point's vicinity through the cross-section of area $\delta F = \delta r \cdot \delta b$ per the time of one period is equal to

$$p(h) \cdot \delta r \cdot \delta n(\delta b) = \frac{\delta F \cdot p(h)}{r \cdot \sin \varphi} \cdot \int_0^{2\pi} p(\Omega) \cdot p(i) \cdot \sin i \cdot d\Omega. \tag{4.20}$$

So, the first problem stated is solved. Now we shall pass to the construction of object's density distribution in azimuth $p(A)$. This density satisfies the relation

$$\int p(A) \cdot dA = 1. \tag{4.21}$$

We shall make use of expression (4.19). The quantity

$$\frac{\delta b}{r \cdot \sin \varphi} \cdot p(\Omega) \cdot p(i) \cdot \sin i \cdot \Delta\Omega = \Phi(\Omega) \cdot \Delta\Omega \tag{4.22}$$

that corresponds to specific values of i, Ω elements and to the discrete increment $\Delta\Omega$, characterises the number of objects (the fraction of δn), which fall inside the δb-vicinity of the point under consideration. All these objects pass through the azimuthal sector $(A, A + \Delta A)$, where

$$\Delta A = \frac{dA}{d\Omega} \cdot \Delta\Omega. \tag{4.23}$$

The derivative is determined from the relation (4.17). It follows from (4.22) and (4.23), that the azimuthal distribution of SO density will be

$$p(A) = k \cdot \Phi(\Omega) / \left(\frac{dA}{d\Omega} \right) = k \frac{\delta b}{r} \cdot \frac{\sin i}{\sin \varphi} \cdot \frac{d\Omega}{dA} p(\Omega) \cdot p(i). \tag{4.24}$$

Here some simplification is applied, that does not have any principal significance. The constant parameters are substituted by a normalised factor k, that is determined from the condition (4.21).

Using the above technique, we developed the program for constructing the distribution of possible directions of velocity vector's tangential component. The distribution $p(i)$ and the set of latitude values φ, for which the distribution $p(A)$ was constructed, served as the initial data. As in the preceding Section, the distribution $p(\Omega)$ was assumed to be uniform.

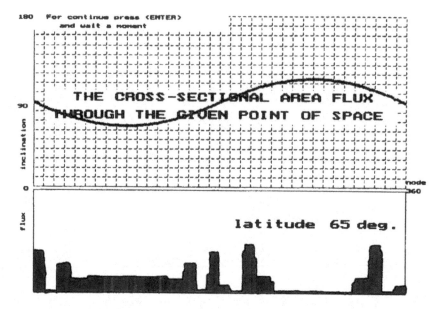

Fig. 4.12. Azimuthal distribution construction for the given point at 65 deg. latitude.

Fig. 4.12 above presents some illustrations for the model. The horizontal axis gives the node longitude values Ω. The upper part of the figure demonstrates the node longitude dependence of inclination i for satellites passing through the given point at 0.65 deg. latitude. The shown values of i and Ω represent the whole set of SOs, which pass through the given point. It is just this curve, along which the curvilinear integral (4.18) is taken. The lower part of the figure presents the $\Phi(\Omega)$ function — the dependence of SO flux of the given type on the node longitude. As follows from expression (4.17), each point of this function corresponds to some certain value of velocity vector azimuth when passing through the given point.

Figs. 4.13 through 4.14 give the normalised distribution of SO velocity vector's azimuth for five chosen points with latitude values of 5, 25, 45, 65 and 85 degrees, respectively. Besides, the first of these figures shows the distribution of azimuths at the equator, which is determined simply from the distribution of inclinations $p(i)$. The latter data indicate, that the most "populated" ranges of inclination are: 65 to 70, 75 to 80 and 90 to 95 degrees (Fig. 4.8). More than 60% of all SOs correspond just to these inclination ranges.

The data obtained show, that the distribution is symmetric with respect to the West-East direction in all versions. At low latitudes (up to 35 deg.) the azimuthal distribution is rather close to the initial distribution of inclinations. One can clearly see here three main maxima corresponding to inclinations in three above-mentioned most frequently applied ranges. As the latitude grows, the region of a maximum (the "petals") become wider, and the maximum themselves shifts to the side of a line of symmetry.

At high latitudes the "petals" are gradually merged and levelled. The distribution tends to uniform one, that is achieved in the pole vicinity.

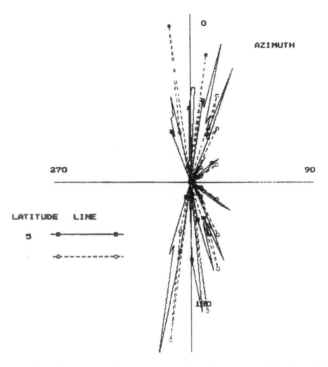

Fig. 4.13. Distribution of velocity vector directions at 0 and 5 deg. latitudes.

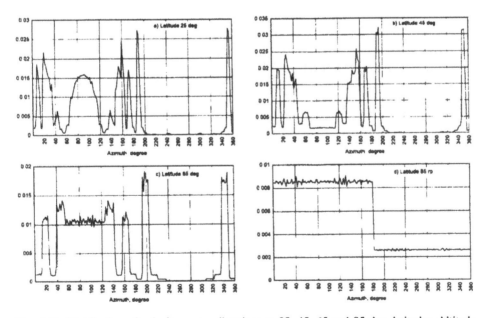

Fig. 4.14. Distribution of velocity vector directions at 25, 45, 65 and 85 deg. latitudes. Altitude range up to 800 km.

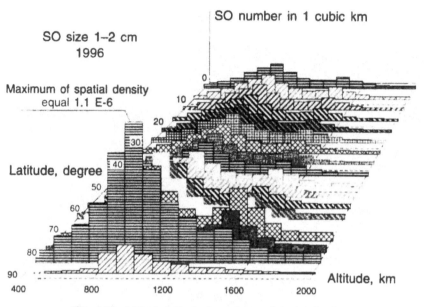

Fig. 4.15. Altitude-latitude dependence of spatial density.

The above results characterise the directions, which are most dangerous for SC. These results essentially differ from the assumption, which was widely applied in the known studies (Kessler, 1978), that the distribution of velocity vector directions is uniform. The obtained results are necessary to determine the distribution of relative velocities for the given SO and space debris particles and, hence, the probability of this SC collision with the other space objects.

4.2.4. Current Spatial Density Distribution

The current distribution of perigee heights and eccentricities of SOs of various size have been constructed by modeling the technogeneous contamination on a preceding time interval. The example of construction of the altitude-latitude distribution of SO concentration is given in Fig. 4.15. The objects of 1–2 cm in size were considered. The absolute maximum of concentration is seen to be achieved at the altitude of 900–1000 km in the latitude range of 80–85 degrees. It equals to $1.10 \cdot 10^{-6}$ objects per 1 cub. km. The second rather prominent (local) maximum lies in the altitude region of 1400–1500 km at latitudes of 70–75 deg. Its value equals 37% of the global maximum. The concentration of objects in the equator region is 20–30% of the maximum concentration at the corresponding altitude. In the altitude range of 400–500 km the concentration of objects is about 4% of the global maximum.

The spatial distribution of the concentration is characterized by function $\rho(h, \varphi)_j$ for various ranges j of SO size depending on the altitude and latitude of a point in the near-

Table 4.4. The maximum values of spatial density of SOs of various size at 1999, km^{-3}

Range No, j	1	2	3	4	5	6	7	8
Size range, cm	0.10–0.25	0.25–0.50	0.50–1.0	1.0–2.0	2.0–4.0	4.0–8.0	8.0–20	>20
ρ_{jmax}	2.49E-4	2.59E-5	6.95E-6	9.64E-7	4.12E-7	1.72E-7	9.39E-8	5.09E-8
Average **m**, kg	0.86E-5	0.58E-4	0.28E-3	0.18E-2	0.010	0.064	0.363	300

Earth space. The accepted ranges of sizes are presented in Table 4.4. The values of function $\rho(h, \varphi)_j$ are presented below as a product

$$\rho(h, \varphi)_j = \rho(h, \varphi)_{j,n} \cdot \rho_{jmax} , \qquad (4.25)$$

where ρ_{jmax} is a maximum value of technogeneous substance spatial density for the j-th range of its sizes, $\rho(h, \varphi)_{j,n}$ is the corresponding normalized value of spatial density.

The maximum values of spatial density for SOs of different sizes are shown in Table 4.4. The example for the value of a normalized spatial density of SOs $\rho(h, \varphi)_{j,n}$ for objects in the size range of 1–2 cm is given in Table 4.5.

The mean value of concentration of technogeneous substance of various size $\bar{\rho}(h)_j$ at the given altitude h is determined by formula

$$\bar{\rho}(h)_j = \sum_k \rho(h, \varphi_k)_j \cdot \cos(\varphi_k) / \sum_k \cos(\varphi_k). \qquad (4.26)$$

These functions are presented in Fig. 4.16. The digits in the right hand side indicate the number of the size range, to which the given curve belongs.

4.3. Collisions of Spacecrafts with Debris Particles Assessment

4.3.1. A Brief Review of the Used Methods

Since sixties the need of increasing spacecrafts (SC) reliability gave rise to the direction, associated with estimating the possibility of their collisions with micrometeoroids. The appropriate techniques have been developed. They were based on the probabilistic approach, which took into account the features of motion of micrometeoroids.

The problem of estimating the possibility of collision of space objects (SO) of artificial origin is relatively new. But only in recent years it became topical and attracted attention of many specialists in different countries. Now this direction is being intensively developed.

The whole set of possible approaches to estimating SO collisions could be sub-divided into two major groups: deterministic and stochastic. The deterministic approach is reliable, when the parameters of motion and the sizes of encountering space objects are very well

Table 4.5. The value of a normalized spatial density for objects in the size range of 1–2 cm

Altitude	$\rho(h, \varphi)_{jn}$ values for various altitudes and latitudes																	
km	2,5	7,5	12,5	17,5	22,5	27,5	32,5	37,5	42,5	47,5	52,5	57,5	62,5	67,5	72,5	77,5	82,5	87,5
450	0.038	0.038	0.038	0.040	0.043	0.049	0.044	0.039	0.040	0.045	0.053	0.055	0.073	0.078	0.083	0.061	0.134	0.011
550	0.083	0.084	0.084	0.089	0.096	0.110	0.099	0.087	0.088	0.099	0.117	0.123	0.162	0.173	0.184	0.135	0.297	0.023
650	0.109	0.110	0.110	0.117	0.126	0.144	0.129	0.114	0.116	0.130	0.153	0.161	0.213	0.227	0.241	0.178	0.389	0.031
750	0.156	0.157	0.157	0.166	0.179	0.205	0.184	0.163	0.164	0.185	0.218	0.229	0.303	0.324	0.343	0.253	0.554	0.044
850	0.176	0.177	0.179	0.186	0.196	0.216	0.208	0.204	0.212	0.234	0.266	0.294	0.372	0.477	0.473	0.432	0.854	0.146
950	0.189	0.190	0.192	0.198	0.207	0.225	0.223	0.226	0.238	0.260	0.292	0.329	0.410	0.552	0.539	0.518	1.000	0.192
1050	0.150	0.151	0.153	0.157	0.164	0.178	0.177	0.181	0.191	0.208	0.234	0.264	0.328	0.448	0.435	0.424	0.813	0.161
1150	0.112	0.113	0.114	0.117	0.123	0.133	0.132	0.133	0.141	0.154	0.173	0.195	0.243	0.326	0.319	0.306	0.591	0.113
1250	0.090	0.090	0.091	0.094	0.099	0.107	0.106	0.107	0.112	0.123	0.138	0.156	0.194	0.260	0.255	0.243	0.471	0.089
1350	0.090	0.091	0.092	0.094	0.099	0.106	0.106	0.109	0.115	0.127	0.142	0.161	0.198	0.257	0.316	0.272	0.386	0.072
1450	0.114	0.114	0.116	0.118	0.123	0.130	0.133	0.141	0.152	0.169	0.184	0.214	0.255	0.330	0.489	0.395	0.393	0.076
1550	0.083	0.083	0.084	0.086	0.089	0.095	0.097	0.102	0.110	0.123	0.134	0.156	0.186	0.238	0.357	0.286	0.278	0.053
1650	0.051	0.051	0.052	0.053	0.055	0.058	0.060	0.062	0.067	0.074	0.082	0.094	0.113	0.143	0.215	0.171	0.168	0.031
1750	0.031	0.031	0.031	0.032	0.033	0.036	0.036	0.037	0.039	0.044	0.048	0.056	0.067	0.085	0.121	0.098	0.107	0.019
1850	0.026	0.026	0.027	0.028	0.029	0.031	0.031	0.032	0.034	0.038	0.042	0.048	0.058	0.073	0.106	0.084	0.090	0.016
1950	0.026	0.026	0.026	0.027	0.028	0.030	0.030	0.031	0.033	0.037	0.041	0.047	0.057	0.069	0.111	0.085	0.073	0.012

Comment: maximum of spatial density equals to 9.64E-7 km^{-3}.

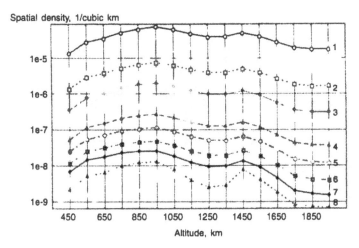

Fig. 4.16. Altitude dependence of average spatial density for SO of various size. 1999

known. In this case, using the motion forecasting algorithms the point of closest approach of a pair of objects is soughed. If the minimum distance between objects occurs to be comparable with their size, the danger of their collision arises. Unfortunately, in most cases the errors of determination of objects' position at the point of closest approach are much larger than their size. By this reason the probability of collision occurs to be small. If it exceeds the value of about 1.0E-4–1.0E-5, the decision can be made about the SC manoeuvre for preventing a possible collision. The number of such precedents does not exceed 3–4 cases.

The technique of calculation the closest approaches of catalogued SOs was developed at the end of sixties. It was used, in particular, in experiments on SO observation by the crew of manned orbital stations. This technique is rather laborious: it requires essential expenditures of computer time because of necessity of multiple addressing to the algorithm of forecasting the SO motion. The most typical example of calculating the probability of collisions on the basis of calculation of closest approaches is given in the paper (Khutorovsky, 1993). The authors called their approach "a direct method". The labour consumption and complexity of a direct method can be evaluated based on the following fact: the searching of pairs of SOs approaching within a day at the approach interval of 50 seconds requires "to pass" through the complex computational procedure up to 10^{11} of pairs of SOs. For each pair of approaching objects the probability of their collision is calculated with account of the accuracy of orbital elements and SO size. The characteristic feature of this calculation is the fact, that the collision probability relates to a particular approach only.

Table 4.6 below presents the histogram of minimum distances between all catalogued

Table 4.6. Number of approaches per one day

ΔR, km	0.1	0.2	0.3	0.5	1.0	2.0	3.0
Number	0.75	3.0	4.6	11.9	47.7	190	426

SOs obtained in the paper (Khutorovsky, 1993). It is seen, that the approaches to distances lower than 100 m are rather rare.

The application of the data on approaches for calculating the probability of collision of some object with others SOs is associated with the necessity of development and application of definite statistical methods of extrapolation of these data. The necessity in their development becomes especially evident, when we consider non-catalogued SOs and future time moments, i.e. the cases, where the necessary initial conditions for forecasting the approaches do not exist.

A typical example of estimating the danger of SO collisions on the statistical approach basis is D. Kessler's technique (Kessler, 1990; Kessler, 1991). The average number of collisions is determined by formula

$$N = S \cdot \rho \cdot V_{rel} \cdot \Delta t , \qquad (4.27)$$

Where S is the cross-section of an object;
ρ is the mean spatial density;
V_{rel} is the mean relative velocity of approaching of two SOs;
Δt is the time interval.

Formula (4.27) is widely applied for estimating the collision danger. Naturally, the calculation results essentially depend on the success of specifying the applied estimations of SO density and average approaching velocity. The further development of the statistical approach is associated not only with the necessity of updating the estimates of density and mean velocity, but also with the necessity of accounting the variety of directions of possible collisions as well as the design (the shape) of a spacecraft. Let's consider this problem in more details, beginning with simplest formulations of the problem of determining the number of collisions of an object with moving particles. By gradual complication of the problem formulation we shall try to reach the acceptable description of a real situation.

4.3.2. Development of the Technique for Collision Probability Evaluation

1. The simplest scheme (Fig. 4.17a). The object is motionless, the velocity of particles, moving with respect to the object, is constant (both in magnitude, and in direction). An obvious example of such a problem is the determination of a number of hailstone particles fallen on the given surface per time unit.

   We designate the object area by S, and the velocity of particles, falling on the object, by V. Apparently, if the density of particles (the number of particles in a volume unit) is ρ, then for the time interval Δt, $\Delta N = S \cdot \rho \cdot V \cdot \Delta t$ particles will fall on a plate. The velocity of increment of an average number of fallen particles will be equal to

$$\frac{dN}{dt} = S \cdot \rho \cdot V . \qquad (4.28)$$

   The quantity $Q = \rho \cdot V$ is called a cross-sectional area flux.

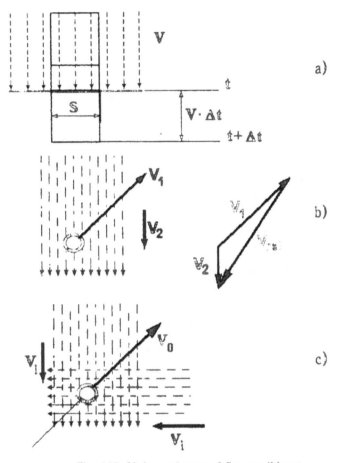

Fig. 4.17. Various schemes of flux conditions:
a) stationary object, only one flux;
b) mobile object, only one flux;
c) mobile object, a few flux.

2. The velocity of particles is constant both in time and in direction, but the object is
   not motionless: it is moving. The difference of the given case from the previous one
   lies in the fact, that the relative velocity of particles $V_{rel} = V_2 - V_1$, rather than their
   absolute velocities, must be used as velocity V (Fig. 4.17b). In calculating the number
   of collisions of an object with particles by formula (4.28) the relative velocity should
   be used as velocity V.

3. Now we complicate the previous formulation. We shall suppose, that there are several
   (a finite number) of directions of particles' motion (Fig. 4.17c). The velocities of each
   of these directions have constant value and constant direction. The values of specific
   flux of particles of the i-th direction will be designated by $Q_i = \rho_i \cdot V_{rel,i}$. The sum of
   all fluxes gives a summary estimate of a specific flux at the given point of space.

Similarly, the total number of collisions of an object will be equal to the sum of a number of collisions with particles of different fluxes

$$\frac{dN}{dt} = S \cdot \sum_i \rho_i \cdot V_{rel,i}.$$

(4.29)

The assumption on the constancy of cross-sectional area S for each of directions of objects approach to the given SC is used here. In the general case this assumption is valid for a spherical surface only.

4. We shall assume, that the number of possible directions of particle fluxes is not restricted, but infinite. Each of directions is determined by two angles in the 3-dimensional space. We shall use the angles of a spherical object-centred coordinate system, namely, azimuth A and elevation angle B.

The density of particles for each of directions will be characterised in the general case by continuous distribution density $p(A, B)$. For more compact presentation of formulas we shall take into account below only the dependence of the distribution density on azimuth A. Such a simplification is not essential, since, first, we can easily transform it to a more full presentation of formulas with account of two arguments and, second, the main flux of space objects has mostly horizontal orientation. This is a consequence of the fact, that an overwhelming majority of objects have small eccentricities. As a result, the density of objects at the given point of space, in a sector, restricted by azimuth values A_i and $A_i + \Delta A$, will be equal to

$$\rho_i = \rho \cdot p(A_i) \cdot \Delta A .$$ (4.30)

After substituting this value into expression (4.29) we obtain

$$\frac{dN}{dt} = S \cdot \rho \cdot \sum_i p(A_i) \cdot V_{rel}(A_i) \cdot \Delta A$$ (4.31)

This expression is used for calculating the collision number N, when the continuous distribution $p(A)$ is implemented as a discrete histogram containing a set of values $p(A_i) \cdot \Delta A$. For correct analytical presentation of formula (4.31) in the continuous case the sum must be substituted by the integral. Then we shall have

$$\frac{dN}{dt} = S \cdot \rho \cdot \int_{A=0}^{2\pi} p(A) \cdot V_{rel}(A) \cdot dA.$$ (4.32)

5. Developing the previous formulation of the problem, we shall take into account the variability of characteristics of a flux of objects at various points of a flight trajectory of the spacecraft under consideration. It was shown in Section 4.2, that both the spatial density of objects ρ, and their azimuthal distribution $p(A)$ very strongly depend on the altitude and latitude of SC. In particular, as the latitude changes, the spatial density

of objects can vary several times, and the azimuthal distribution varies from a strongly "cut up" one (at the equator) up to uniform one (above the Earth's poles).

Since an integration of the equations of motion of SC the altitude and latitude of the SC are functions of time, the variability of objects' characteristics can be easily taken into account based on the equation (4.32), by including time t into the list of variables appearing in the right-hand side of this equation. As a result, we can write

$$\frac{dN}{dt} = S \cdot \rho(t) \cdot \int_{A=0}^{2\pi} p(t, A) \cdot V_{rel}(t, A) \cdot dA. \quad (4.33)$$

The quantity $Q(t)$ in the right-hand side of this expression, i. e.

$$Q(t) = \rho(t) \cdot \int_{A=0}^{2\pi} p(t, A) \cdot V_{rel}(t, A) \cdot dA \quad (4.34)$$

has a meaning of a specific relative flux of objects at a given point of orbit. And the integral has a meaning of the mean relative velocity of SO. It follows from expression (4.34), that the quantity

$$\Delta Q(t) = \rho(t) \cdot p(t, A) \cdot V_{rel}(t, A) \cdot \Delta A \quad (4.35)$$

has a meaning of a specific flux of objects through the azimuthal sector $(A, A + \Delta A)$ at a current point of a trajectory. Summing-up and averaging these estimates over the interval of one revolution is a basis for constructing the law for the relative velocity directions and magnitude distribution at collision.

A characteristic feature of the technique described above lies in the fact, that, on one hand, it takes into account in details the variability of the SO flux as a function of orbital elements of the SC under consideration and of its position in the near-Earth space (NES), and, on the other hand, it allows to calculate rather simply the mean expected number of SC collisions with space debris and statistical characteristics of the relative velocity.

The possibility of a rather simple application of the proposed technique for practical calculations of characteristics of possible SC collisions with space debris exists under the conditions, that variable functions, $\rho(t)$, $p(t, A)$ and $V(t, A)$, appearing in the right-hand side of expression (4.33), are known. The technique and some results of construction of these quantities were discussed in the previous Section 4.2.

The variables appearing in the right-hand side of expression (4.33) are almost-periodic functions. The basic periodic component of these functions has a period equal to the revolution period of the SC under consideration. The long-periodic and secular components are associated with a slow evolution of SC orbital elements and with a slow change of the degree of NES contamination with technogenous objects. In view of above considerations, it is meaningful to perform the averaging of SO flux through the unit cross-section per 1 revolution (for the time period equal to the SC period T).

This mean value is calculated by formula

$$\overline{Q} = \frac{1}{T} \cdot \int\limits_{t=0}^{T} \rho(t) \cdot \int\limits_{A=0}^{2\pi} p(t,A) V_{rel}(t,A) \cdot dA \cdot dt. \qquad (4.36)$$

If the mean value of flux $\overline{Q}$ is determined, then a rather accurate estimate of a number of SC collisions with space debris on a large time interval $t - t_0$ can be calculated by a very simple formula, which is similar to the formula (4.27)

$$N = S \cdot \overline{Q} \cdot (t - t_0). \qquad (4.37)$$

It is meaningful to perform calculations of the values of mean cross-sectional flux $\overline{Q}$ for various types of SC, as well as for current and future times for convenience of performing simple engineering calculations on the basis of the formula (4.37).

All methods of calculating the expected average number of object collisions with debris particles are based on the supposition, that these particles are small-sized, i.e. their size is much smaller than the size of an object under consideration. This assumption is reflected in the above formulas by including only the mean cross-sectional area S of the object under consideration. To take into account the size of particles (if it is comparable with object's size), the further improvement of the technique is necessary.

6. We shall assume, that the space debris can have various sizes including those, which can not be neglected. Possible sizes of particles will be characterised by the distribution density $p(d)$ of their mean diameter d. Taking into account the distribution density $p(d)$ allows to introduce into the equation (4.33) modifications, accounting for the variability of particles' sizes. In this case the mean diameter of the SC under consideration will be designated by D. As a result, we get

$$\frac{dN}{dt} = \left[\frac{\pi}{4} \int\limits_{d} (D+d)^2 \cdot p(d) dd \right] \cdot \rho(t) \int\limits_{A=0}^{2\pi} p(t,A) \cdot V_{rel}(t,A) \cdot dA. \qquad (4.38)$$

Here the expression in square brackets has a meaning of the total calculated area of SC and space debris under consideration. In all other respects this expression coincides with (4.33). The idea of using the sum $(D + d)$ as the basic parameter, that influences the probability of collision of objects of various size has been used in a series of works (Khutorovsky, 1993, Rossi, 1992). Such an approach is justified in the cases, when the shape of colliding particles can be approximated by a sphere. Otherwise, when the shape of SC cross-sections strongly varies as particles arrive at the SC from different directions (that could be the case in the presence of large solar batteries), the further sophistication of the technique is required. This problem still has to be solved in future.

To calculate the probability (P) of at least one collision with the known value of N many authors recommend to apply the Poisson distribution

$P(N) = 1 - \exp(-N)$. (4.39)

However, it is obvious, that at small $N \ll 1$ the value of probability only slightly differs from N, i. e. $P \approx N$. In this case there is no necessity in applying formula (4.39), and the collision probability may be taken to be equal to the mean number of expected collisions N. There is no major sense in applying formula (4.39) for large N values as well, since as N increases, the probability P tends to 1. This means, that the formula (4.39) is suitable for calculation of a single collision only, which is not correct in estimating collisions with large number of small-sized technogeneous particles, providing the possibility of multiple collisions within considerable time frames. Thus, in this case the quantity N itself (4.38) provides a better estimation.

4.3.3. Characteristics of the Relative Flux of SOs

Table 4.7 presents the results of application of the proposed technique for estimating the space debris flux with respect to some typical SC orbits. 8 ranges of object sizes were considered with boundary values of 0.1, 0.25, 0.5, 1.0, 2.0, 4.0, 8.0 and larger than 20 cm. Particle sizes from 0.10 to 0.25 cm correspond to range No. 1, and those larger than 20 cm–to the range No. 8.

Typical SC orbits were assumed to be circular. The reference values for altitudes and inclinations of orbits were the following ones:

— for altitudes h: 400, 600, 800, 1000, 1200, 1400 km.
— for inclinations i: 55, 65, 75, 85 and 95°.

The table data show, that all considered arguments (the space debris size range, the altitude and inclination of SC orbit) have essential influence on the characteristics under consideration. For given values of inclinations the corresponding flux densities differ not more than 1.7 times. The influence of altitude is more essential: for indicated altitudes the corresponding estimates differ an order of magnitude approximately. The influence of sizes is most significant: the corresponding estimates differ 6000–7000 times in this case.

Table 4.8 presents the data on the mean velocity of SC collisions with space debris on various orbits differing in altitude and inclination. The particle size dependence is not presented, since it is weakly exhibited. These data indicate, that the values of mean velocity of possible collisions vary within the limits from 10.8 to 13.2 km/s depending on the inclination and altitude of SC orbit. The inclination has greatest influence on these estimates. The remaining factors are insignificant.

Figures 4.18 and 4.19 present the obtained distributions of magnitude and direction of possible collisions for SC with altitude of 950 km and inclinations of 55, 75 and 95°. The direction of possible collision is characterised by the deviation (clockwise) of the relative velocity vector for the approaching SO from the direction of motion of the SC.

SPACE DEBRIS: HAZARD EVALUATION AND MITIGATION

Table 4.7. Cross-sectional area flux with respect to different circular SC orbits, 1999

Inclination	Size Range cm	Values of cross-sectional area flux ($m^{-2} \cdot year^{-1}$) for different altitudes (km)					
		400	600	800	1000	1200	1400
55°	0.1–0.25	0.28E-2	0.75E-2	0.13E-1	0.17E-1	0.11E-2	0.11E-1
	0.25–0.5	0.29E-3	0.80E-3	0.14E-2	0.18E-2	0.12E-2	0.11E-2
	0.5–1.0	0.81E-4	0.22E-3	0.39E-3	0.48E-3	0.29E-3	0.28E-3
	1.0–2.0	0.12E-4	0.33E-4	0.54E-4	0.65E-4	0.38E-4	0.36E-4
	2.0–4.0	0.57E-5	0.15E-4	0.24E-4	0.27E-4	0.15E-4	0.14E-4
	4.0–8.0	0.26E-5	0.69E-5	0.10E-4	0.11E-4	0.60E-5	0.57E-4
	8.0–20	0.16E-5	0.40E-5	0.56E-5	0.58E-5	0.28E-5	0.29E-5
	>20	0.41E-6	0.13E-5	0.24E-5	0.27E-5	0.83E-6	0.12E-5
65°	0.1–0.25	0.30E-2	0.81E-2	0.14E-1	0.18E-1	0.11E-1	0.11E-1
	0.25–0.5	0.32E-3	0.87E-3	0.15E-2	0.19E-2	0.12E-2	0.12E-2
	0.5–1.0	0.88E-4	0.24E-3	0.42E-3	0.50E-3	0.31E-3	0.29E-3
	1.0–2.0	0.13E-4	0.36E-4	0.59E-4	0.68E-4	0.40E-4	0.38E-4
	2.0–4.0	0.63E-5	0.17E-4	0.26E-4	0.29E-4	0.16E-4	0.15E-4
	4.0–8.0	0.29E-5	0.75E-5	0.11E-4	0.12E-4	0.62E-5	0.59E-5
	8.0–20	0.17E-5	0.44E-5	0.62E-5	0.61E-5	0.30E-5	0.30E-5
	>20	0.45E-6	0.14E-5	0.26E-5	0.28E-5	0.87E-6	0.12E-5
75°	0.1–0.25	0.37E-2	0.10E-1	0.18E-1	0.23E-1	0.14E-1	0.15E-1
	0.25–0.5	0.40E-3	0.11E-2	0.19E-2	0.24E-2	0.15E-2	0.16E-2
	0.5–1.0	0.11E-3	0.30E-3	0.54E-3	0.64E-3	0.39E-3	0.40E-3
	1.0–2.0	0.16E-4	0.44E-4	0.75E-4	0.87E-4	0.51E-4	0.53E-4
	2.0–4.0	0.78E-5	0.21E-4	0.33E-4	0.37E-4	0.21E-4	0.21E-4
	4.0–8.0	0.36E-5	0.93E-5	0.14E-4	0.15E-4	0.80E-5	0.85E-5
	8.0–20	0.21E-5	0.55E-5	0.79E-5	0.79E-5	0.39E-5	0.44E-5
	>20	0.56E-6	0.17E-5	0.34E-5	0.37E-5	0.11E-5	0.18E-5
85°	0.1–0.25	0.41E-2	0.11E-1	0.20E-1	0.27E-1	0.17E-1	0.16E-1
	0.25–0.5	0.43E-3	0.12E-2	0.22E-2	0.29E-2	0.18E-2	0.17E-2
	0.5–1.0	0.12E-3	0.33E-3	0.60E-3	0.77E-3	0.47E-3	0.43E-3
	1.0–2.0	0.18E-4	0.49E-4	0.85E-4	0.10E-3	0.60E-4	0.56E-4
	2.0–4.0	0.85E-5	0.23E-4	0.37E-4	0.44E-4	0.25E-4	0.22E-4
	4.0–8.0	0.39E-5	0.10E-4	0.16E-4	0.18E-4	0.10E-4	0.87E-7
	8.0–20	0.23E-5	0.60E-5	0.85E-5	0.95E-5	0.46E-5	0.44E-5
	>20	0.61E-6	0.19E-5	0.38E-5	0.44E-5	0.14E-5	0.17E-5
95°	0.1–0.25	0.46E-2	0.12E-1	0.23E-1	0.28E-1	0.17E-1	0.16E-1
	0.25–0.5	0.49E-3	0.13E-2	0.24E-2	0.29E-2	0.18E-2	0.17E-2
	0.5–1.0	0.13E-3	0.37E-3	0.68E-3	0.75E-3	0.46E-3	0.43E-3
	1.0–2.0	0.20E-4	0.54E-4	0.95E-4	0.10E-3	0.59E-4	0.55E-4
	2.0–4.0	0.95E-5	0.26E-4	0.42E-4	0.43E-4	0.24E-4	0.22E-4
	4.0–8.0	0.44E-5	0.11E-4	0.18E-4	0.18E-4	0.94E-5	0.86E-5
	8.0–20	0.26E-5	0.67E-5	0.99E-5	0.93E-5	0.45E-5	0.43E-5
	>20	0.69E-6	0.21E-5	0.43E-5	0.43E-5	0.13E-5	0.17E-5

Table 4.8. Mean velocity of SC collision with space debris

Inclination degrees	Velocity values (km/s) for SC with different altitudes (km)					
	400	600	800	1000	1200	1400
55	10,8	10,7	10,6	10,9	11,2	11,2
65	11,3	11,2	11,1	11,7	11,6	11,5
75	12,2	12,2	12,1	12,0	12,4	12,3
85	12,5	12,4	12,3	12,9	12,8	12,7
95	13,2	13,1	13,0	12,7	12,6	12,6

As the data of Fig. 4.18 show, the increase of SC orbit inclination causes the growth of the fraction of possible collisions with SOs moving on encountering courses (the "frontal" collisions). This is reflected in the distribution of the collision velocity value (Fig. 4.19): the relative fraction of collisions with high velocities grows.

The data presented above reflect general regularities in distributions of collision velocities values and directions. At the same time, it should be noted, that these data do not cover the whole diversity of possible conditions of SC interaction with space debris particles. For elliptical orbits the character of the considered distributions drastically changes — the asymmetry of possible directions of collisions arises, the altitude and perigee argument of the orbit become essentially influential factors. Therefore, in the general case, for a more thorough analysis of conditions of possible collisions the use of a special model is required.

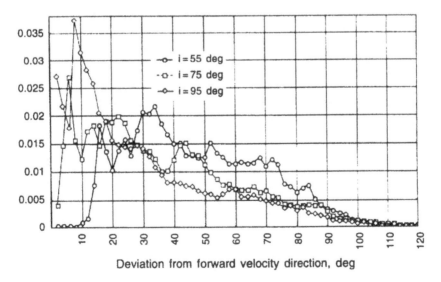

Fig. 4.18. Angular distribution of SD relative velocity with respect to SC of various inclinations, h = 400 km.

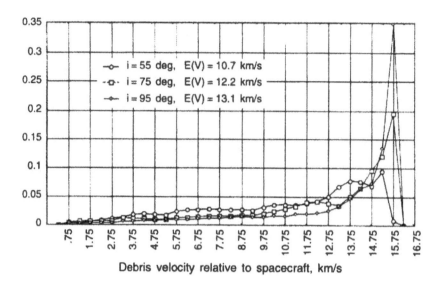

Fig. 4.19. Collision velocity distribution versus SC inclination, h = 400 km.

4.3.4. The Probability of Mutual Collisions for a Group of Objects

The study and estimation of probabilities of mutual collisions of objects belonging to different groups — large-size (catalogued), medium-size (from 1 up to 20 cm) and small-size (for example, from 0.1 up to 1 cm), etc. is of considerable interest. For this purpose it is convenient to express the spatial density of particles sizing larger than arbitrary number d as a product of some dimensionless factor $k(d)$ by the spatial density of particles sizing larger than some specified value d_0:

$$\rho(d, t) = k(d) \cdot \rho(d_0, t) \tag{4.40}$$

Here coefficient $k(d)$ is supposed to be independent of time. We designate the derivative of coefficient $k(d)$ as $f(d) = dk(d)/dd$. Then the spatial density of particles in the range of (d_1, d_2) can be expressed as follows:

$$\rho(d_1, d_2, t) = -\int_{d_1}^{d_2} f(d) \cdot dd \cdot \rho(d_0, t) = [k(d_1) - k(d_2)] \cdot \rho(d_0, t). \tag{4.41}$$

Here it is taken into account, that the values of $k(d)$ decrease as the size d increases. The generalisation of the expression (4.38) for this case is the equation

$$\frac{dN(D, d_1, d_2)}{dt} = \left[-\frac{\pi}{4} \int_{d}^{} (D + d)^2 \cdot f(d) \cdot dd \right] \cdot \rho(d_0, t) \cdot V_{rel}(t). \tag{4.42}$$

The convenience of application of approximation (4.40) is, in fact, that the expression in square brackets turned to be time-independent. Therefore, there is no necessity in calculating the integral in square brackets at each time step of integration. It is sufficient to calculate it only once.

$$F_d = \left[-\frac{\pi}{4} \int_{d_1}^{d_2} (D+d)^2 \cdot f(d) \cdot dd \right]. \tag{4.43}$$

Besides, at a proper selection of function $f(d)$ this integral can be taken analytically. With using designation (4.36) the solution of equation (4.42) over the interval $t - t_0$ can be written in the following form:

$$N(D, d_1, d_2) = F_d \cdot \overline{Q}(d_0, t) \cdot (t - t_0). \tag{4.44}$$

Here the cross-sectional area flux, appearing in the right-hand side, is calculated in accordance with the integral (4.36). In this case the spatial density $\rho(d_0, t)$ of particles sizing larger than specified value d_0 is used as the required spatial density. It is convenient to choose the lower boundary of size of catalogued objects as the quantity d_0. It equals to 20 cm approximately. In this case $Q(d_0, t)$ is the cross-sectional area flux of catalogued SOs with respect to some SC moving over the specified orbit. Naturally, the quantity $Q(d_0, t)$ depends on orbital parameters — altitude, inclination, eccentricity. Numerous investigations showed, that the influence of altitude was the most essential. In this study we neglect other arguments due to the fact, that with changing inclination the cross-sectional area flux varies not more, than by 15–20%, as well as due to the fact, that the majority of orbits are close to circular ones.

Thus, for some particular SO (with size D) we obtained the estimate of a number of collisions $N(D, d_1, d_2)$ with all others SOs sizing in the range of (d_1, d_2). For transition from estimate $N(D, d_1, d_2)$ to the estimate of a mean number of collisions of a group of objects having size in the range of (D_1, D_2), which are in some altitude region $(h, h + \Delta h)$, with all SOs having size in the range of (d_1, d_2) (this estimate is designated below as $N(h, h + \Delta h)_{Dd}$), it is necessary to sum up estimates $N(D, d_1, d_2)$ for all SOs of the given size being in the given altitude range. In this case for a sufficiently small Δh all estimates $Q(d_0, t)$ could be considered identical. As a result, we obtain the following estimate:

$$N(h, h + \Delta h)_{Dd} = \left[\sum_j F_d(D_j) \right] \cdot \overline{Q}(d_0, t_0') \cdot (t - t_0). \tag{4.45}$$

Here, in square brackets, the summation of F_d values, calculated according to expression (4.43), is carried out over all SOs having size in the range (D_1, D_2), for which the altitudes lie in the range $(h, h + \Delta h)$. The expression in square brackets can be written in a slightly different manner, if we introduce into consideration the total number of objects $n(D > D_j)$ with size $(D > D_j)$ and the functional dependence of a number of objects on their size, which is similar to (4.40):

$$n(D_j > D) = k(D) \cdot n(h, h + \Delta h)_{cat} .$$ (4.46)

Here function $k(D)$ is the same, as that applied above in expression (4.40), and $n(h, h + \Delta h)_{cat}$ is the number of catalogued objects in the altitude range $(h, h + \Delta h)$. Using formula (4.46), we can easily determine the number of objects having size in the range $(D, D + dD)$. This number is equal to $f(D) \cdot dD \cdot n(h, h + \Delta h)_{cat}$. With regard to this consideration, the summation in (4.45) is substituted by the integral. As a result, we obtain

$$\left[\sum_j F_d(D_j) \right] \approx \left[\frac{\pi}{4} \int_{D_1}^{D_2} \int_{d_1}^{d_2} (D+d)^2 \cdot f(d) \cdot dd \cdot f(D) \cdot dD \right] \cdot n(h, h + \Delta h)_{cat}.$$ (4.47)

The expression in square brackets in the right-hand side of (4.47) will be denoted by F_{Dd}. Then expression (4.45) takes the form

$$N(h, h + \Delta h)_{Dd} = F_{Dd} \cdot n(h, h + \Delta h)_{cat} \cdot \overline{Q}(d_0, t_0) \cdot (t - t_0).$$ (4.48)

where F_{Dd} is calculated by formula

$$F_{Dd} = \left[\frac{\pi}{4} \int_{D_1}^{D_2} \int_{d_1}^{d_2} (D+d)^2 \cdot f(d) \cdot dd \cdot f(D) \cdot dD \right].$$ (4.49)

The calculation of estimates (4.49) for various groups of SOs and the comparison of the obtained values with each other provides the relative estimates of probability of collisions of SOs inside various groups and between the groups. In performing particular calculations we have used the assumption, that the relation between the logarithm of density and logarithm of size inside each group is linear. The discrete values of the function $k(d)$ at points $d_j, j = 1, ..., m$ will be considered as the known ones. Here m is the number of groups, into which all the regarded sizes of objects are sub-divided. Then the function $k(d)$ will have in the range (d_j, d_{j+1}) the following form

$$k(d) = k(d_{j+1}) \cdot \left(\frac{d_{j+1}}{d} \right)^{x_j} = \frac{C_j}{d^{x_j}},$$ (4.50)

where

$$x_j = \frac{\ln(k(d_j) / k(d_{j+1}))}{\ln(d_{j+1} / d_j)}.$$ (4.51)

It follows from (4.50), that function $f(d)$ looks as follows:

Table 4.9. Matrix of the F_{Dd} values (sq. m) for SO of three size groups

Size range, cm	0.01–0.1	0.1–20	> 20
0.01–0.1	790–1900	400–800	1800–3600
0.1–20	400–800	3.9–10.0	6.5–20.0
> 20	1800–3600	6.5–20.0	0.60–1.0

$$f(d) = -x \cdot \frac{C_j}{d^{x_j+1}}. \tag{4.52}$$

The substitution of (4.52) into (4.49) results in the following formula for calculating quantity F_{Dd}:

$$F_{Dd} = \frac{\pi}{4} \cdot C_j \cdot C_i \cdot x_j \cdot x_i \cdot$$

$$\left\{ \left[\frac{D^{3-x_j}}{3-x_j} \right]_{D_1}^{D_2} \cdot \left[\frac{d^{1-x_i}}{1-x_i} \right]_{d_1}^{d_2} + 2 \left[\frac{D^{2-x_j}}{2-x_j} \right]_{D_1}^{D_2} \cdot \left[\frac{d^{2-x_i}}{2-x_i} \right]_{d_1}^{d_2} + \left[\frac{D^{1-x_j}}{1-x_j} \right]_{D_1}^{D_2} \cdot \left[\frac{d^{3-x_i}}{3-x_i} \right]_{d_1}^{d_2} \right\} \tag{4.53}$$

An example is shown below wherein all SOs sizing larger than 0.1 ñì are sub-divided into 3 groups with respect to size. The characteristics of these groups are the following:

j	1	2	3
d_j, cm	0.10	1.0	20
$k(d)$	10600	44	1
C_j	0.00758	0.1309	0.030
x_j	2.382	1.263	2.177

The results of calculations are given in Table 4.9. Here possible scattering of initial data ($k(d)$ values) is taken into account. The results seem to be quite interesting and important.

First, expression (4.48) characterises a spatial distribution of the probability (mean number) of collisions.

Second, the table data show, that as the SO size under consideration decreases, the number of collisions increases some orders of magnitude — both inside the given group and between the groups. In particular, the number of collisions of dangerous fragments sizing 1 to 20 cm with catalogued SOs is 10.9 + 25 times larger, than the number of possible collisions of catalogued SOs between each other. In this case the number of possible mutual collisions of particles from the second group 4.8–15.0 times differs from the number of mutual collisions of catalogued objects.

The number of possible collisions of catalogued SOs between each other per one year has the order of 0.01 now (Nazarenko, 1996). It follows from the given data, that the number of collisions of particles from the 1st group of sizes (from 0.1 to 1.0 cm) between each other has a value of about several tens. The number of collisions of 1st group objects

Table 4.10. Matrix of the F_{Dd} values (sq. m) for SO of four size groups

Size range, cm	0.01–0.1	0.1–10	10–20	>20
0.01–0.1	1123	310	248	3600
0.1–10	310	3.7	1.3	15.8
10–20	248	1.3	0.14	0.67
>20	3600	15.8	0.67	0.95

with the objects from 2nd and 3rd groups has the same order of magnitude. It can be supposed that in such collisions a great number of smaller fragments is formed, which velocities are essentially lower than the initial ones. These particles will no longer be orbital, i.e. the Earth's satellites. While becoming aerosols, they, basically, sediment slowly on the Earth. A part of them could mix up with the atmosphere.

Thus, alongside with the natural process of NES self-purification due to particles slowing down in the atmosphere, another self-purification process takes place — due to mutual collisions of small-size fragments of space debris between each other.

The estimate $F_{Dd} = 0.60 + 1.0$ m^2 for collisions between catalogued objects (inside the 3rd group) is of particular interest. Using relation $F_{DD} = 2 \cdot (\pi/4) \cdot \overline{\overline{D}}^2$, we can determine the mean size $\overline{\overline{D}}$ inside the group. In particular, for catalogued SOs (of 3rd group) we obtain $\overline{\overline{D}} = 0.62$–$0.80$ m. This estimate is in a rather good agreement with the published data on catalogued objects' size.

In some papers it was supposed, that the dangerous collisions (from the point of view of contribution into the so-called cascade effect) are those of objects sizing larger than 10 cm. The number of such SOs is about 2.5 times greater than the number of catalogued objects. By this reason a more detailed sub-division of SOs (into 4 groups) is considered below. The characteristics of these groups are:

j	1	2	3	4
d_j, cm	0.10	1.0	10	20
$k(d)$	10600	44	2.5	1
C_j	0.00758	0.1309	0.1191	0.0460
x_j	2.382	1.263	1.322	1.912

For this case matrix F_{Dd} is presented in Table 4.10. It is obvious from the expression (4.48), that the total number of collisions of objects of 3rd and 4th groups between each other per 1 year is equal to the product of quantities:

$$\overline{\overline{F}} = F_{3+4} = (F_{33} + F_{34} + F_{43} + F_{44})/2 = (0.14 + 0.67 + 0.67 + 0.95)/2 = 1.21 \text{ sq. m,}$$

and the sum   (4.54)

$$Sum = \sum_j S(h_j) = \sum_j n(h_j, h_j + \Delta h)_{cut} \cdot Q(d_0, t_0)_j.$$

Table 4.11. Values of $\overline{\overline{F}}$ (sq. m) for space debris of various size

SD size, cm	≥ 0.01	≥ 0.10	≥ 10	≥ 20
$\overline{\overline{F}}$	4740	21.4	1.2	0.475

If we considered the 4th group only, then the total number of collisions of objects inside the 4th group would be proportional to the quantity $\overline{\overline{F}} = F_{4+4} = (F_{44})/2 = 0.475$ sq. m. This quantity has a meaning of some mean area $\overline{\overline{F}}$ of objects under consideration.

The values of $S(h_j)$ and Sum do not depend on groups of objects under consideration. The total number of collisions N_{sum} of objects of various sizes is equal to the product of quantities $\overline{\overline{F}}$ and Sum:

$$N_{sum} = \overline{\overline{F}} \cdot Sum . \qquad (4.55)$$

In considering the number of collisions of various groups of objects only the value of the mean area $\overline{\overline{F}}$ will vary. Table 4.11 presents the $\overline{\overline{F}}$ values for objects from groups with numbers $j \geq 1$, $j \geq 2$, etc. These data show, that at transition from the group of sizes with $j \geq 3$ to the group of sizes with $j \geq 2$ the number of collisions increases 17.7 times. At the same time, the number of collisions for $j \geq 1$ exceeds 221 times the number of collisions inside the groups and between the groups with $j \geq 2$. This is a typical feature of the size dependence of a number of objects. For $d < 1$ cm this dependence is more strong as compared with the objects sizing $d > 1$ cm (the values of x_j are different).

The altitude distribution of a number of collisions is characterised by the values of $S(h_j)$. Figure 4.20 presents the normalised values for the altitude distribution of a collision number:

$$n(h_j)_S = S(h_j)/Sum . \qquad (4.56)$$

Here Sum is the number of collisions per one year for catalogued objects.

Another two altitude functions are presented in Fig. 4.20: the distributions of values of the mean spatial density of catalogued objects

$$n(h_j)_\rho = \overline{\overline{\rho}}(h_j) / \left(\sum_j \overline{\overline{\rho}}(h_j) \right) \qquad (4.57)$$

and values of a square of density

$$n(h_j)_{\rho\rho} = \overline{\overline{\rho}}(h_j)^2 / \left(\sum_j \overline{\overline{\rho}}(h_j)^2 \right) . \qquad (4.58)$$

These data clearly indicate, that the altitude distributions of a collision number and a square of spatial density virtually coincide. This property is expedient to be used for

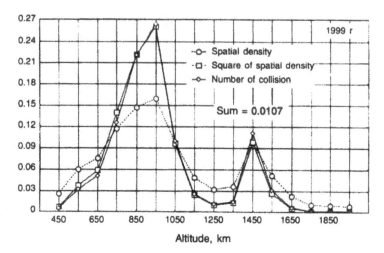

Fig. 4.20. Altitude distribution: a) spatial density (23); b) square of spatial density (24); and c) number of collision (22).

saving computer time during the evolution modelling with regard to the consequences of possible collisions.

In summary of the unit we shall consider outcomes of calculations of a matrix F_{Dd} for a more detailed partition of sizes (Tables 4.12 and 4.13). From these data it is not so easy to receive values of the matrix F_{Dd} at a more rasping partition. For this purpose it

Table 4.12. Subdivision of space debris sizes into eight groups

Group No (j)	1	2	3	4	5	6	7	8
d_j, cm	0.100	0.215	0.464	1.00	2.15	4.64	10.0	21.5
$k(d)$	10600	1703	274	44	16.7	6.33	2.40	1.0
C_j	0.000723	0.000786	0.000758	0.1295	0.1317	0.1310	0.1724	0.0505
x_j	2.389	2.375	2.382	1.266	1.261	1.263	1.144	1.942

Table 4.13. Matrix of the F_{Dd} values (sq. m) for SO of eight size groups

j/i	1	2	3	4	5	6	7	8
1	458	184	95	47.6	76.5	129	211	3258
2	184	54.8	22.0	9.19	13.5	21.7	34.6	524.7
3	95	22.0	6.60	2.10	2.61	3.82	5.82	85.0
4	47.6	9.19	2.10	0.48	0.45	0.56	0.76	10.24
5	76.5	13.5	2.61	0.45	0.32	0.30	0.35	4.01
6	129	21.7	3.82	0.56	0.30	0.21	0.19	1.63
7	211	34.6	5.82	0.76	0.35	0.19	0.13	0.67
8	3258	524.7	85.0	10.24	4.01	1.63	0.67	1.03

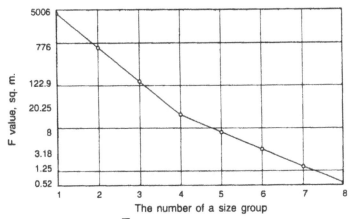

Fig. 4.21. Value $\overline{\overline{F}}$ dependence versus a size group number.

is necessary to sum up the values in the relevant cells. For example, the total of all values F_{Dd} at $j = 1,2,3$ and $i = 1,2,3$ is peer 1121. Above for this group of sizes the rating 1123 is adduced. The small variance (long of percent) is explained by influence of rounding at the representation of reference values $k(d)$.

Below in the table and in Fig. 4.21 the values of $\overline{\overline{F}}$ for particles from groups with the numbers $j \geq 1$, $j \geq 2$, etc. are assembled.

Groups	$j \geq 1$	$j \geq 2$	$j \geq 3$	$j \geq 4$	$j \geq 5$	$j \geq 6$	$j \geq 7$	$j = 8$
$\overline{\overline{F}}$	5006	776	122.9	20.25	8.00	3.18	1.25	0.52

The data analysis shows that in a logarithmic scale the relation $\overline{\overline{F}}(d_j)$ consists of two linear distributions. In the interval of sizes (0.1–1.0) cm it looks like

$$\overline{\overline{F}}(d_j) = 5006 \cdot (0.1/d_j)^{2.393},\qquad (4.59)$$

and in the interval of sizes (1–21.5) cm:

$$\overline{\overline{F}}(d_j) = 20.25 \cdot (1/d_j)^{1.209}.\qquad (4.60)$$

The parameters of exponentiation 2.39 and 1.21 included in these functions practically coincide with the reduced relevant ratings for $j = 1,2,3$ and $j = 4,5,6$.

4.4. The Account of Shape and Orientation of Typical Spacecraft Modules

4.4.1. Aerodynamic Analogy

The atmospheric drag is the main disturbing factor for SC flying up to altitudes of about 500–800 km. The drag force monotonously lowers the orbit height and, as a result, the

SC enters dense layers of the atmosphere and ultimately burns down. The SC drag force in the atmosphere is determined by the relationship:

$$F_x = \frac{1}{2} \cdot C_x \cdot S \cdot \rho \cdot V_{rel}^2. \qquad (4.61)$$

Here C_x is the non-dimensional aerodynamic drag coefficient, S is the characteristic area of SC, ρ is the atmospheric density, V_{rel} is the relative SC velocity with respect to air.

Below is considered the generalisation of the expression (4.37) for the case, when the configuration of SC differs from a spherical one. One can easily see that the right-hand sides of expressions (4.61) and (4.27) have very much in common. In both cases they contain the values of SC characteristic area (S), particles density ρ and relative SC/particles velocity. Such a similar character of expressions is due to the fact, that in both cases the characteristics of particles interaction with the surface are determined.

The main difference between expressions under consideration lies in the fact, that they relate to different characteristics. The drag force in the atmosphere is estimated in the first case and the space debris flux — in the second. These distinctions are revealed in quadratic and linear dependences of estimated characteristics on the relative velocity.

The above common features do not prove the analogy yet. Let us generalise the above-mentioned technique of evaluating the mean number of SC collisions with small-sized debris to the case, when the SC configuration differs from a spherical one. In so doing we shall apply the technique of taking into account the SC configuration used in aero-dynamics.

We break the whole outer surface of SC (F) into elementary sections of area dF. The total space debris flux through the elementary area is a combination of the particles flows having different directions of relative velocity (angle A). Let us introduce the angle β between the external normal to the elementary area and arbitrary direction of relative velocity vector of the incident space debris flux. Then the total particles flux through the elementary area will be equal to

$$\frac{dN_{dF}}{dt} = dP(dF) = dF \cdot \rho(t) \cdot \int_A \cos\beta \cdot p(t,A) \cdot V_{rel}(t,A) \cdot dA. \qquad (4.62)$$

In determining the debris flux through the outer SC surface the integral must include only those terms, for which $\cos\beta > 0$.

Summing up the flux (4.62) over all elementary areas provides the unknown estimate of space debris flux through the surface of SC under consideration. The corresponding equation will take form:

$$\frac{dN}{dt} = P = \int_F dP(dF) = \rho(t) \int_F \int_A \cos\beta \cdot p(t,a) \cdot V_{rel}(t,A) \cdot dA \cdot dF. \qquad (4.63)$$

Here the integral in the right-hand side is taken over the whole SC surface. The condition of obtaining the proper estimation of a total flux is the absence of the outer surface shading.

Formally, the shading condition implies, that there exists some element on the surface, the external direction of flux to which intersects the other surface element of the given SC.

By analogy with expression (4.61), we shall present the total flux P as a product of characteristic area S, spatial density $\rho(t)$, mean collisions velocity at the given point and some coefficient C_N. We obtain the following relationship:

$$\frac{dN}{dt} = P = C_N \cdot S \cdot \rho(t) \cdot \bar{V}_{rel}(t). \tag{4.64}$$

Here the mean collision velocity is determined as

$$\bar{V}_{rel}(t) = \int_A p(t,A) \cdot V_{rel}(t,A) \cdot dA, \tag{4.65}$$

and coefficient C_N is defined according to (4.63) and (4.64) by the expression

$$C_N = \frac{P}{S \cdot \rho \cdot \bar{V}_{rel}} = \frac{\displaystyle\iint_{F\,A} \cos\beta \cdot p(t,A) \cdot V_{rel}(t,A) \cdot dA \cdot dF}{S \cdot \displaystyle\int_A p(t,A) \cdot V_{rel}(t,A) \cdot dA}. \tag{4.66}$$

Coefficient C_N is non-dimensional. It takes into account the SC configuration. In particular case, for the spherical SC and at $S = \pi \cdot R^2$, the value of this coefficient equals to 1. Thus, expression (4.64) is the generalisation of formulas (4.27) and (4.37), that takes into account the influence of SC shape on the number of collisions with space debris. This coefficient is an analogue of the coefficient C_x, which is widely applied in the aerodynamics.

By analogy with the approach applied in astrodynamics, for convenience of performing simple engineering calculations based on formula (4.37), it makes sense to calculate coefficient C_N and mean values of space debris' relative velocity for various SC flight conditions. Since the mean values are more stable as compared with instantaneous ones, there arises a possibility of their preliminary determination for rather wide conditions and tabulation.

As it is seen from the expression (4.9), the values of coefficient C_N depend upon the distribution of directions of a space debris flux $p(t,A)$ and possible values of relative collision velocity $V_{rel}(t,A)$. The character of these dependences for space debris essentially differs from their analogues in the aerodynamics.

The counter flux directions principally differ from each other. As it was mentioned above and shown in paper (Nazarenko, 1997), the distribution of directions of a counter flux of technogeneous debris particles is planar and "highly rugged". Besides, it strongly depends on the latitude: it coincides with the distribution of inclinations at the equator and is uniform over the poles. Examples of these distributions are shown in Fig. 4.19.

The value of possible velocity of SC collision with space debris, $V_{rel}(t,A)$, varies within very wide limits: from 0 up to $2V_\tau$, where V_τ is the tangential component of SC velocity.

With an accuracy, sufficient for practical calculations, the relative collision velocity is determined by the relationship:

$$V_{rel}(t, A) \approx 2 \cdot V_r(t) \cdot \cos(A/2).$$ (4.67)

Thus, the analogy between aerodynamics and the estimation of danger of various SCs collision with technogeneous space debris is not absolute. There are some essential differences, which cause a necessity in special investigations and calculations with the purpose of evaluating the influence of the SC configuration on the collisions danger.

4.4.2. Determining the Coefficient C_N for Typical SC Structure Components

A special program was developed for solution of this problem. The technique is based on the materials derived in this Section and on the results of application of the SDPA model. Four typical SC structure configurations were considered: a cylinder, a cone, a panel and a hemisphere. For the given orbital elements the equation (4.63) has been integrated within the limits of one orbital revolution.

The surface of a typical component is divided into elementary areas. For each of them one determines the collision probability with a due account of all possible directions of particles' incident flux, and then one creates the histogram of a cosine of the angle of relative velocity vector deviation from the normal to the surface. The resulting data are determined by integrating over all elementary areas.

Let us consider in more detail the method of specifying the configuration and orientation of typical blocks. *The configuration* is specified by means of indicators for cylinder, cone, panel and hemisphere, respectively. *The dimensions* of a cylinder are characterised by diameter and length, for cone — by a major diameter, length and minor diameter, for panel — by its area, and for hemisphere — by its diameter.

The orientation is specified by means of two angles α and β in the movable Cartesian co-ordinate system connected with SC. Axis OX is aligned with the radius — vector, and axis OY lies in the orbital plane in the motion direction (i.e. it is aligned with the tangential velocity component). Axis OZ supplements the co-ordinate system up to right- hand one.

The angle α is an analogue of azimuth. It is counted off in the horizontal plane (YOZ) clockwise from the axis OZ. Angle β is an analogue of the angle of site. This is the angle between the specified direction and horizontal plane. The axis orientation is specified for a cylinder, cone and hemisphere, and the orientation of a normal to the surface is specified for a planar element. All these angles are shown in Fig. 4.22.

In order to analyse in details the interaction of a flux of particles with the surface of a typical SC component, one should specify the co-ordinate of an arbitrary elementary area of its surface. This is necessary, in particular, for all non-planar configurations, since in this case the characteristics of particles fluxes will be different for different sections of the surface. This specifying is performed by means of angle φ (see Fig. 4.22), which characterises the position of a generatrix of the body of revolution (cylinder, cone or hemisphere) relative to the line of intersection of a rear end wall with a horizontal plane

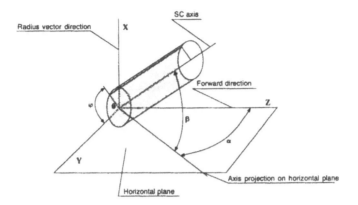

Fig. 4.22. Orientation of typical SC structure components.

(axis y1). At the initial orientation of a typical component ($\alpha = 0$, $\beta = 0$) axis y1 coincides with axis OY. At arbitrary orientation of a component this axis is shifted with respect to the OY axis by angle α.

For typical components in the form of bodies of revolution the parameters of their collisions with space debris are calculated *for corresponding external surfaces only.* The end surfaces are not considered here. If necessary, they should be specified as some individual planar elements.

The data on the C_N coefficient values for all possible variants of orientation of SC type elements with an inclination in the range 50–60 degrees are shown in the Fig. 4.23 and

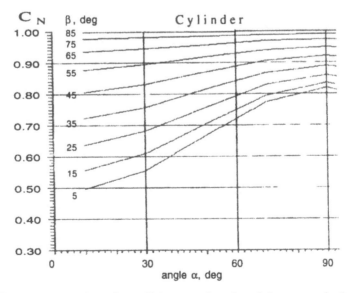

Fig. 4.23. The dependence of the C_N coefficient on orientation of the spacecraft of a cylindrical shape.

and in the Tables 4.14–4.18 as well. On approximating the β angle to 90 degrees (i.e. to a perpendicular to the flux plane) the values of the C_N coefficient became independent of the α angle (the azimuth). In this case they approach 1 for rotation bodies and 0 for a panel.

On carrying out the calculations for inclinations, different from 50–60 degrees, it is recommended to apply presented values of the C_N coefficient for obtaining approximated estimations. For more precise calculations it is recommended to use the computing SDPA model (Nazarenko, 1997).

4.4.3. Examples of Collision Probability Determination

Let us consider determining the number of impacts of the simplified model of «Space station» with space debris sizing larger than 1 cm. This simplified test model was examined

Table 4.14. Cone (semi-angle = 15°)

B°	C_N values for different cone orientation, A°								
	10	30	50	70	90	110	130	150	170
5.0	0,9319	0,9448	0,9499	0,9250	0,8380	0,6915	0,5109	0,3527	0,2576
15.0	0,9430	0,9519	0,9580	0,9332	0,8503	0,7068	0,5323	0,3779	0,2893
25.0	0,9813	0,9854	0,9833	0,9522	0,8710	0,7397	0,5837	0,4467	0,3679
35.0	1,0251	1,0238	1,0125	0,9750	0,8979	0,7829	0,6512	0,5367	0,4706
45.0	1,0606	1,0547	1,0366	0,9961	0,9262	0,8302	0,7246	0,6341	0,5820
55.0	1,0815	1,0726	1,0512	1,0121	0,9528	0,8774	0,7980	0,7313	0,6931
65.0	1,0839	1,0743	1,0537	1,0204	0,9750	0,9212	0,8670	0,8228	0,7977
75.0	1,0659	1,0582	1,0427	1,0197	0,9908	0,9589	0,9283	0,9041	0,8906
85.0	1,0271	1,0239	1,0177	1,0091	0,9990	0,9886	0,9791	0,9720	0,9681

Table 4.15. Cone (semi-angle = 45°)

B°	C_N values for different cone orientation, A°								
	10	30	50	70	90	110	130	150	170
5.0	2,5805	2,3836	2,0286	1,5568	1,0874	0,6816	0,3838	0,1679	0,0620
15.0	2,5087	2,3218	1,9821	1,5293	1,0762	0,6808	0,3874	0,1736	0,0668
25.0	2,3692	2,2023	1,8914	1,4776	1,0557	0,6815	0,3953	0,1868	0,0780
35.0	2,1721	2,0335	1,7636	1,4079	1,0311	0,6885	0,4117	0,2121	0,1013
45.0	1,9445	1,8321	1,6168	1,3303	1,0136	0,7094	0,4501	0,2602	0,1570
55.0	1,7403	1,6516	1,4838	1,2588	1,0052	0,7553	0,5374	0,3766	0,2905
65.0	1,5379	1,4733	1,3517	1,1878	1,0020	0,8168	0,6544	0,5339	0,4696
75.0	1,3276	1,2883	1,2143	1,1143	1,0008	0,8871	0,7873	0,7130	0,6734
85.0	1,1102	1,0970	1,0720	1,0385	1,0002	0,9619	0,9283	0,9032	0,8899

Table 4.16. Cone (semi-angle = 75°)

B°	C_N values for different cone orientation, $A°$								
	10	30	50	70	90	110	130	150	170
5.0	9.4065	8.3841	6.6571	4.7034	2.8934	1.4320	0.5132	0.1119	0.0060
15.0	9.1212	8.1324	6.4596	4.5684	2.8106	1.3965	0.5024	0.1116	0.0063
25.0	8.5594	7.6367	6.0712	4.3029	2.6481	1.3269	0.4818	0.1110	0.0071
35.0	7.7384	6.9128	5.5053	3.9162	2.4134	1.2266	0.4534	0.1109	0.0086
45.0	6.6839	5.9838	4.7821	3.4218	2.1173	1.1004	0.4213	0.1124	0.0114
55.0	5.4302	4.8806	3.9300	2.8387	1.7784	0.9561	0.3930	0.1182	0.0178
65.0	4.0227	3.6455	2.9902	2.1988	1.4305	0.8121	0.3847	0.1368	0.0349
75.0	2.5620	2.3726	2.0273	1.5920	1.1425	0.7434	0.4326	0.2246	0.1201
85.0	1.4455	1.3918	1.2921	1.1606	1.0155	0.8750	0.7553	0.6687	0.6233

Table 4.17. Panel

B°	C_N values for different panel orientation, $A°$								
	10	30	50	70	90	110	130	150	170
5.0	0.8018	0.7107	0.5616	0.3898	0.2421	0.1107	0.0375	0.0051	0.0000
15.0	0.7774	0.6891	0.5445	0.3779	0.2348	0.1073	0.0364	0.0049	0.0000
25.0	0.7294	0.6465	0.5109	0.3546	0.2203	0.1007	0.0341	0.0046	0.0000
35.0	0.6593	0.5844	0.4618	0.3205	0.1991	0.0910	0.0309	0.0042	0.0000
45.0	0.5691	0.5044	0.3986	0.2767	0.1718	0.0786	0.0266	0.0036	0.0000
55.0	0.4616	0.4092	0.3233	0.2244	0.1394	0.0637	0.0216	0.0029	0.0000
65.0	0.3401	0.3015	0.2382	0.1654	0.1027	0.0469	0.0159	0.0021	0.0000
75.0	0.2083	0.1846	0.1459	0.1013	0.0629	0.0288	0.0098	0.0013	0.0000
85.0	0.0701	0.0622	0.0491	0.0341	0.0212	0.0097	0.0033	0.0004	0.0000

Table 4.18. Hemisphere

B°	C_N values for different hemisphere orientation, $A°$								
	10	30	50	70	90	110	130	150	170
5.0	1.8021	1.7060	1.5246	1.2801	1.0016	0.7232	0.4780	0.2959	0.1986
15.0	1.7778	1.6845	1.5087	1.2716	1.0016	0.7316	0.4939	0.3173	0.2230
25.0	1.7298	1.6423	1.4773	1.2548	1.0015	0.7481	0.5251	0.3594	0.2709
35.0	1.6596	1.5805	1.4314	1.2303	1.0013	0.7723	0.5708	0.4210	0.3410
45.0	1.5694	1.5011	1.3724	1.1988	1.0011	0.8035	0.6295	0.5002	0.4312
55.0	1.4618	1.4065	1.3020	1.1612	1.0009	0.8406	0.6994	0.5946	0.5386
65.0	1.3403	1.2995	1.2225	1.1188	1.0007	0.8825	0.7785	0.7013	0.6600
75.0	1.2084	1.1834	1.1363	1.0727	1.0004	0.9280	0.8644	0.8170	0.7918
85.0	1.0702	1.0617	1.0459	1.0245	1.0001	0.9757	0.9543	0.9384	0.9299

Table 4.19. Calculation of impact number per 1 year

No	Module	Surface	S, m²	C_N	$C_N \cdot S \cdot \bar{Q}$	NASA data
1	Box	Front	1	0.631	0.618E-5	
		left side	0.215	0.346	0.073E-5	
		right side	0.215	0.346	0.073E-5	
		Total	–	–	**0.76 E-5**	**0.44 E-5**
2	Starboard	Cylinder	1	0.637	0.624E-5	
		Face	0.785	0.346	0.266E-5	
		Total	–	–	**0.89 E-5**	**0.73 E-5**
3	Port	Cylinder	2	0.637	1.248T-5	
		Face	0.785	0.346	0.266E-5	
		Total	–	–	**1.51 E-5**	**1.18 E-5**
4	Trailing	Cylinder	3	0.692	**2.03 T-5**	**1.87 E-5**
5	Earth	Cylinder	1	1.000	**0.98 E-5**	**0.98 E-5**
		Total sum			**6.17 E-5**	**5.30 E-5**

in the work (Kessler *et al.*, 1996). The simplified analogue of SS consists of a cube with the side of 1 m and four connected to it cylinders with diameters of 1 m. Lengths of these cylinders with various axes directions are: opposite to the velocity vector (Trailing) 3 m; on the binormal (starboard — to the right) 1 m; to the opposite side (port — to the left) 2 m; aside the Earth center 1 m.

The parameters of the circular orbit are: the altitude of 400 km, the inclination of 51.6 degree. The data on space debris environment date to 2002. The value of average specific flux of particles sizing larger than 1 cm according to SDPA model is $\bar{Q} = 0.98 \cdot 10^{-5}$ m^{-2}·year^{-1}. Table 4.19 contains the data on the average number of impacts per one year (Number Impacts), obtained by developed methods and on the base of NASA 96 model (Christiansen, 1998).

The comparison of the SPDA and NASA models indicates the following:

- Estimations of an expected total impacts number are rather similar. The divergence does not exceed 16%.
- There is some divergence in correlation of fluxes through different elements. For example, fluxes through the Earth component coincide, fluxes through the Box component in the SDPA model are 1.7 times higher. All these differences are apparently due to divergence of distributions of the values and direction of particles velocities.

In the Fig. 4.24 statistical distributions of the velocity values for possible impacts calculated according to SPDA and ORDEM96 (Kessler, 1996) models are presented.

The data indicate the presence of a considerable difference between calculations results for the two examined models. It is seen that the distribution provided by the SPDA model is shifted to the right with respect to the NASA distribution, i.e. impacts with greater velocities are more frequent. For the SPDA model the average value of the impact velocity was 10.6 km/sec, and for the ORDEM96 model it was 8.8 km/sec (by 17% less).

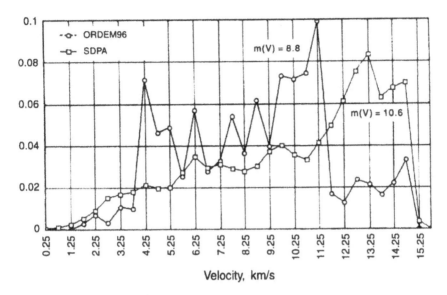

Fig. 4.24. Comparison of the collision velocity distributions for ISS.

The reasons for the divergence of the velocity distributions are apparently due to applying different hypotheses of the eccentricities distribution. This problem was examined in details in the report (Nazarenko, 1997). The NASA model assumes only two values for eccentricities: 0 and 0.5. In the SPDA model breaking up the range of possible eccentricities values is more detailed and consists of 8 «boxes».

Let us examine in more details possible basic reasons for these distinctions. The output file of the ORDEM96 model contains data on the characteristics of the relative velocity in the following form:

Velocity V, km/sec	Angle A(V), degrees	p(V)
0.25	89.0	0.00168
0.75	86.6	0.00131
1.25	0.00	0.00000
1.75	83.2	0.01377
2.25	81.5	0.02608
2.75	79.0	0.01377
...	...	...

Each value of the relative velocity V (with a set step ΔV) is characterized by: $A(V)$ — the declination of the direction of the relative velocity from the direction of the SC velocity vector and the probability $p(V)$ of the velocity getting into the vicinity ($V \pm \Delta V/2$) of the given value. Enumerated characteristics are developed separately for space debris at circular and elliptical orbits.

Analogous characteristics in the SPDA model have another form:

A, degrees	$V(A)$, km/sec	$p(A)$
1	15.86	0.00011
3	15.84	0.00023
5	15.81	0.00023
7	15.75	0.00039
9	15.68	0.00041
...	...	...

Each inclination of relative velocity direction (A) from the direction of SC velocity vector (with set step ΔA) is characterized by: the values of relative velocity $V(A)$ and the probability $p(A)$ of getting of the relative velocity direction into the vicinity ($A \pm \Delta A/2$) of the given value. Enumerated characteristics are general for all space debris.

On estimating the probability of impacts and penetrations in the ORDEM96 and SDPA models the cycles are organized based on different arguments: on V *and* A respectively. It may seem that these two approaches are equivalent, as corresponding distribution densities are connected by the formula:

$$p(A) \cdot dA = p(V) \cdot dV. \tag{4.68}$$

However, in fact the precision of results of each approach is significantly different. Let us compare the functions $V(A)$, developed by each model. They are shown in Figs. 4.25 and 4.26.

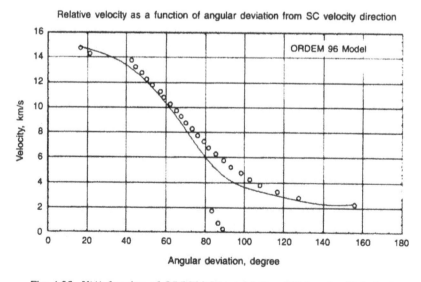

Fig. 4.25. $V(A)$ function of ORDEM 96 model ($h = 400$ km, $i = 51.6$ deg.)

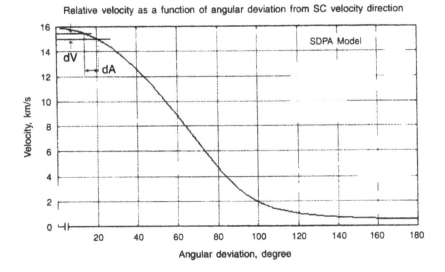

Fig. 4.26. V(A) function of SDPA model (h = 400 km, i = 51.6 deg.)

From these data one can learn the following properties of the functions:

1. The general character of the dependence of the relative velocity on the impact direction is the same for both models.
2. The data of the ORDEM96 model is not smooth. In the vicinity of impact directions that are characterised by the angles of 0 and 90 degrees, some points fall out off the aggregated line.
3. In the ORDEM96 model a number of points in the vicinity of low angular deflections (under 40 degrees) is insufficient for correct description of impacts in this area of flying-up angles.
4. In the vicinity of flying-up angles larger than 90 degrees, the values of the relative velocity in the ORDEM96 are essentially higher than that in the SDPA model.

Enumerated in items 2 and 3 features are of fundamental importance. They indicate the presence of three concrete sources of errors in the ORDEM96 model:

1. The presence of uncertainties in the initial data modeling, that become apparent in the «non-smoothness» of $V(A)$ distribution.
2. An intensification of these errors on transition to the distribution $p(A)$ in the vicinity of small flying-up angles owing to low values of the derivatives dV/dA in this area. The Fig. 4.26 illustrates this effect. Applying the data of the ORDEM96 model for the impacts after-effects analysis this intensifications becomes apparent objectively, irrespective of either the distribution $p(A)$ is constructed or not.
3. A scanty points number in the vicinity of small flying-up angles.

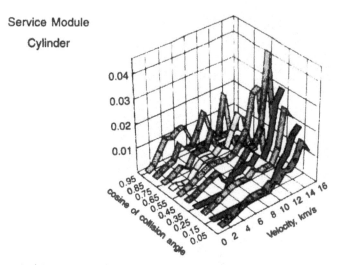

Fig. 4.27. Bivariate distribution of relative velocity and cosine of collision angle.

In that way the difference of presented in the Fig. 4.24 velocity distributions of possible impacts may be explained by insufficiently adequate description provided by the ORDEM96 model for a great number of impacts, which directions are similar to frontal ones.

In conclusion of this part we give an example of constructing the two-dimensional distribution of the impact velocity and the cosine of the velocity angular deflection from the normal to the surface (Fig. 4.27). This distribution is relative to the cylindrical surface of the auxiliary modulus of the International Space Station. The orientation of the modulus axis coincides with the velocity vector. Such distribution is used for estimation of the hazard of SC side penetration. From the figure data it can be seen that in such a case impacts, which directions are similar to the normal, take place not very often. Besides those impacts have low relative velocity (below 6 km/sec). In that way the auxiliary modulus seems to be in favorable conditions from the standpoint of a side penetration danger.

4.5. Forecast of Space Debris Environment

4.5.1. General Characteristics of the Forecasting Algorithm

The algorithm is intended for altitude distribution forecasting of SO of various sizes, both catalogued (sizing larger than 10–30 cm) and non-catalogued (sizing smaller than 10–30 cm) ones. The algorithm application area is: the altitude range 400–2000 km, sizes range larger than 1 mm.

The upper limit of the altitude range has been chosen because forecasting the evolution of objects distribution in the examined area is the most difficult task, and also because the overwhelming majority of SO (approximately 90%) have now the perigee altitude lower than 2000 km. At higher altitudes the self-purification process of space does not go, so there is a monotonous growth of a number of SO in accordance with their generating intensity. When we determine the lower limit of the altitude range we have to take into account the exponential growth of SO braking with lowering the altitude. The strong atmospheric influence at the altitudes below 400 km results in a rather short existence of SO in this region and in stabilization of the pollution process there. During the next several tens and even hundreds of years it is not expected to have significant changes of situation there. So it has been chosen $h_{min} = 400$ km.

The limit of the examined range of SO sizes (1 mm) has been chosen under the circumstances that objects of just that size are of the greatest danger for working SC. Besides among the smaller-sized SO the particles of natural origin (micro-meteorites) are predominant. The technology of preparing the initial data is described in the Part 4.1. Below it is listed briefly.

1. Sub-division of all the SO into 8 groups by sizes ranges (Table 4.1) is performed.
2. Two-dimensional histograms are created for the distribution of perigee altitudes and ballistic coefficients for non-catalogued SO of the j-th sizes range at the initial moment: $p(h, S)_j$ is a number of SO in each of the «boxes» of a two-dimensional area. The altitude step is 100 km, the scale of the values for ballistic coefficients is: 1.5, 0.5, 0.15, 0.05, 0.015, 0.005 m^2/kg. This histogram is relative to the forecast initial time moment (1997, 1998, 1999 and so on). It gets renewed annually on the base of the forecasting modeling of evolution of the technogenous contamination.
3. Normalized histograms of eccentricities distribution of SO of various sizes $p(e)_j$ are created. The scale for the values of eccentricities is: 0.0025, 0.0075, 0.025, 0.07, 0.15, 0.3, 0.5 and 0.7. An example of such histograms is present in the Fig. 4.2.
4. The data on the generating intensity of new objects and on their characteristics (Fig. 4.5) is introduced.
5. The forecasting range in years is set.

The initial data could be taken from corresponding files and it could be also edited and set in a dialogue regime. The statement of the problem and the initial equations were discussed in the Part 4.1. An analytical solution of the initial equations doesn't seem possible. So the numerical-analytic method was applied.

Calculating the distribution evolution of the SO number versus altitude the following factors were taken into account:

- Slowing down of SO in the atmosphere at the altitudes under 2000 km;
- dividing of all the SO into groups differing by sizes d_j, the eccentricity values e and the ballistic coefficients S;
- the initial distribution of SO of various types versus altitude;
- the intensity of generating of SO of various types as a result of new launches and orbital explosions ($q(t, h, \chi)$ is an increase of the number of SO at different altitudes per time unit);

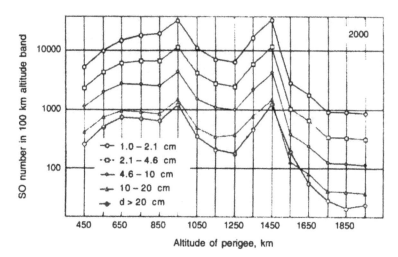

Fig. 4.28. Distributions of perigee altitudes for SOs of various sizes.

• non-stationary character of the considered factors (periodical variations of the atmospheric density caused by the solar activity changes during 11-year cycle; the intensity of new launches variations, etc.).

4.5.2. Initial Environment and Conditions of the Forecast

The initial conditions for the forecast relate to the year 2000. They were generated by means of SDPA Model by modeling the technogeneous contamination on the previous time interval. The range of altitudes from 400 to 2000 km was considered. The data on contamination by particles sizing larger than 1 cm were taken into account.

Figure 4.28 presents the distribution of perigee altitudes of space objects (SO) of various sizes in 2000. The whole range of sizes was divided into 5 intervals. SOs larger than 20 cm in size were supposed to be the catalogued ones. In accordance with the initial data approved by IADC, all these objects were divided into 4 groups: payloads, upper stages, mission-related objects, fragmentation debris. Thus, the total number of standard sizes of SOs equals $(5 - 1) + 4 = 8$. The distributions of eccentricities for these 8 types of SOs are given in Table 4.20.

Besides, the statistical distributions of inclinations and ballistic factors of objects of various types were generated. The Table 4.21 presents the data on a number of objects of various types in 2000. The total number of objects sizing larger than 1 cm was 282 thousands.

The debris environment forecast conditions were generated on the basis of the analysis of technogenous contamination on a previous interval. These conditions are given in Table 4.22.

The altitude distributions of each of standard sizes of SOs at the time of their generation were formed. Five different scenarios were used for the debris environment forecasts.

Table 4.20. Distribution p(e) for SD of various sizes (cm)

e	non-catalogued				catalogued (>20 cm)			
	1–2.1	2.1–4.6	4.6–10	10–20	fragment	related	upper st.	Payload
0.0010	0.065	0.076	0.092	0.1103	0.140	0.18	0.18	0.40
0.0035	0.112	0.131	0.158	0.1969	0.250	0.38	0.38	0.33
0.0075	0.134	0.157	0.184	0.2189	0.254	0.14	0.14	0.09
0.0150	0.194	0.217	0.251	0.2590	0.235	0.11	0.11	0.03
0.0400	0.361	0.339	0.270	0.1974	0.119	0.03	0.03	0.02
0.0800	0.090	0.062	0.039	0.0166	0.002	0.03	0.03	0.02
0.2000	0.044	0.018	0.006	0.0009	0.000	0.03	0.03	0.03
0.4000	0.000	0.000	0.000	0.0000	0.000	0.10	0.10	0.08

Table 4.21. Number of objects of various sizes (cm)

Non-catalogued SOs				Catalogued SOs				
1–2.1	2.1–4.6	4.6–10	10–20	>20 cm	fragments	related	upper st.	Payload
174113	66451	26254	8985	6730	2427	2079	808	1416

Table 4.22. Conditions for the technogeneous contamination forecast

Characteristic	Value
Initial moment of time	Year 2000
Forecast interval	50 years
Altitude range under consideration	400–2000 km
Space debris size under consideration	larger than 1 cm
Solar activity	11-year cycle
The data on the nominal annual surplus of a number of various SOs	
Spacecraft (SC)	52.0
Launch vehicles (LV)	60.4
Mission-related objects sizing > 20 cm	128
Nominal annual number of explosions	3
Average number of catalogued SOs per 1 explosion	36.9
The ratio of a number of SOs sizing larger than 1 cm to the nominal surplus of a number of catalogued SOs	50

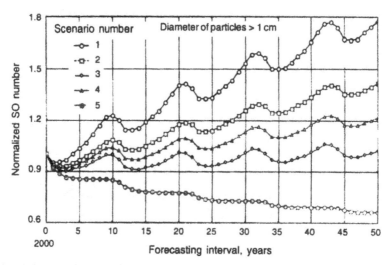

Fig. 4.29. Variations of the number of particles sizing larger than 1 cm for various scenarios of technical policy.

Those scenarios characterized the influence of various strategies of prevention the catastrophic NES contamination.

Scenario 1   Business as usual: the contamination intensity is the same as during the last 10 years

Scenario 2   Scenario I plus elimination of mission-related objects: no mission-related objects

Scenario 3   Scenario 1 plus avoidance of explosions: no fragmentation

Scenario 4   2-fold reduction of a number of launches and associated mission- related objects and breakup fragments

Scenario 5   All measures of Scenarios 2, 3 and 4 are applied simultaneously

4.5.3. Example of a Debris Environment Forecast

Figure 4.29 illustrates the results of forecasting the number of objects under the conditions of application of each of scenarios considered above. The ratio of a current number of objects to their corresponding number at the initial moment is used here.

The obtained results testify the following:

1. At preserving the current technical policy (Scenario 1) *the basic law of change of space debris quantity is its monotonous growth.* In 50 years the number of particles sizing larger than 1 cm will increase 1.8 times and will reach 500 thousands approximately. A rather small periodic component is caused by the 11-year cycle of solar activity and has no principal importance.

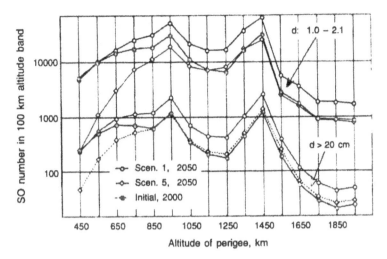

Fig. 4.30. Comparison of altitude distributions of space debris at the initial time moment and in 2050.

2. The considered ways of limitation of the space debris formation intensity (Scenarios 2, 3 and 4) result in essential decrease of NES contamination rates. *The greatest effect is achieved in the case of prevention of satellite explosions* (Scenario 3). The prevention of formation of mission related fragments (Scenario 2) gives the lowest efficiency.

3. *Each of the considered measures separately does not prevent the monotonous growth of technogenous NES contamination.*

4. *Only combined application of various ways of lowering the intensity of space debris formation (Scenario 5) results in the total decrease of its quantity* (by 25–30%).

The obtained estimates are rather significant. They testify that the problem of prevention of monotonous NES contamination is quite complicated, which requires a complex approach to its solution based on application of all possible ways of lowering the intensity of space debris formation.

The analysis of evolution of the altitude distribution of space objects is of a certain interest. It allows to find the areas, where the application of measures on mitigating the technogeneous contamination is the most urgent task. The basic results of this analysis are present in Fig. 4.30. This figure gives the comparison of space debris versus altitude distributions at the initial time moment and in 2050.

Three altitude distributions were considered for each of these sizes: the initial one in 2000 and two distributions in 2050. One of them corresponds to the current contamination intensity (Scenario 1), and the other one — to application of complex measures on mitigating the contamination (Scenario 5). These data allow to make the following conclusions.

1. As compared to the altitude distribution, with keeping the current contamination

intensity the application of complex measures on mitigating the contamination *will result in lowering the space debris quantity at all altitudes*, especially lower than 1000 km.

2. The simultaneous application of the considered strategies for lowering the space debris formation intensity (Scenario 5) *does not result in essential reduction of technogenous contamination at altitudes higher than 900–1100 km. The monotonous contamination by objects larger than 10 cm in size will continue in this region.*

   The last conclusion testifies to the necessity of taking additional measures directed at mitigating the technogeneous contamination at the altitudes higher than 1000 km.

CHAPTER 5

GEOSTATIONARY ORBIT POLLUTION AND ITS LONG-TERM EVOLUTION

T. Yasaka

Accumulation of space debris and collision hazards in Geosynchronous orbit are discussed in the present chapter. Collision breakup models and results for GEO debris environment evolution evaluation are presented. Necessary preventive and mitigation measures are analyzed with account of peculiarities of Geosynchronous orbit population.

5.1. Objects Accumulation and Collisions Hazard in GEO

5.1.1. Objects Accumulation in GEO

The geostationary orbit (GEO) is a unique natural resource, which provides an ideal location for Earth oriented mission satellites. A large number of communication and observation satellites have been injected into this orbit since early stage of space utilization. Importance of this orbit can never be over emphasized by mentioning the significant influence of these satellites brought forth to welfare of human life. Launches into this orbit have been one of the largest drivers of space development because of clear advantage of GEO systems over terrestrial systems in relaying and collecting information that private sectors find interesting as well as government agencies do. This unique natural resource should be kept as useful as it is today for many generations to come. However, this orbit is quite fragile against pollution by space debris. Lack of atmospheric drag makes orbital lifetime of debris infinitely long. Therefore, once the orbit is polluted by a numerous number of debris, GEO looses its usefulness forever, unless an effective action is taken

113

to sweep them out. Unfortunately, there is no effective way known to lessen the number of small debris in GEO.

In February 1963, SYNCOM 1 was launched by the US. This was the first satellite, which was intended to be put into a geo-synchronous orbit. Although the mission was unsuccessful due to a failure during apogee motor burn, recent catalogue includes this satellite, and the orbital period is almost the same as that of the earth rotation, and the inclination is 33 degrees. It was only five years and three months after the launch of the first artificial satellite. Following SYNCOM 2, which successfully performed communication from geo-synchronous altitude in an inclined orbit, the first geostationary satellite SYNCOM 3 was launched in August 1964 (Flight International, 1964). It was kept stationary over the Pacific Ocean and relayed Tokyo Olympics to the other side of the globe. Since then, launches into this orbit became quite common, and as a result, the number of objects increased continuously until over 700 objects were accumulated in the vicinity of GEO by 422 launches as of the end of 1998.

Geostationary satellites are normally put into an elliptical transfer orbit by launchers and the satellites fire a motor at the apogee point to transfer themselves into the stationary orbit. During their operational life span, they maintain the orbit on the equatorial plane at a specified longitude. At the end of life, or just prior to exhaustion of orbit control propellant, they are commanded to move into a higher circular orbit by using residual propellant. Those which do not perform this re-orbit maneuver at the end of life are left uncontrolled under natural perturbing forces. The orbital inclination starts to change cyclically, with the maximum of 15 degrees and the period of 56 years. The radius also changes but its fluctuation remains within a small distance of less than 50 km. The orbit raised satellites are also subjected to natural perturbation forces, but the orbits are normally high enough so that they do not come down to the altitude at which other satellites operate. In this way, the GEO is intended to be used by a limited number of actively controlled operational satellites. In actuality, about 200 uncontrolled satellites that ceased to operate in the past without a re-orbiting maneuver are orbiting at the geo-synchronous altitude. In addition, there are many rocket motors in the vicinity of GEO. Proton and some types of old Titan make direct insertion of payload into GEO, and their final stage motors are separated at an altitude very close to GEO. Many meteorological satellites separate their apogee motors after insertion into the circular orbit. More than 150 of these boosters are now in orbit. In Fig. 5.1, the number growth of objects in and near GEO is shown. Typical motions of these objects within a day are illustrated in Fig. 5.2.

To make a more detailed analysis of the presently orbiting objects, a complete object data set provided by NASDA through its DOANATS system in October 1998 is used. This data set is based on US catalogue, and both are potentially identical. Out of 8382 object records, 909 objects are found of interest to GEO environment. They have perigee radius of more than 40000 km, but they exclude those of high inclinations which are mostly Molniya type objects and those with extremely high perigees. These objects are then classified into several groups as shown in Fig. 5.3, in order that each group represents specific feature of objects. Perigee radius of 20000 km is chosen to separate "Elliptical" and "Near Circular" orbit. 199 objects belonging to "Elliptical" include most of upper stage boosters that launched satellites into geo-transfer orbit. "Near Circular" objects are then

Object Accumulation

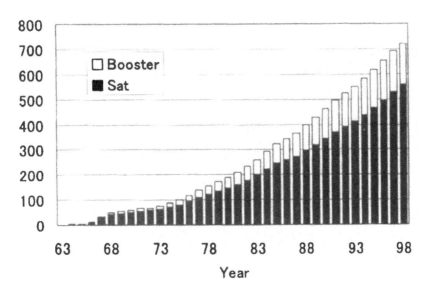

Fig. 5.1. Object accumulation in GEO and its vicinity.

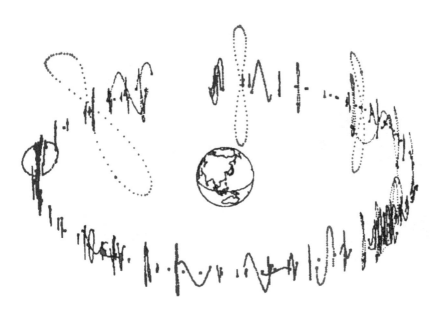

Fig. 5.2. Typical motion of objects in geo-synchronous altitude.

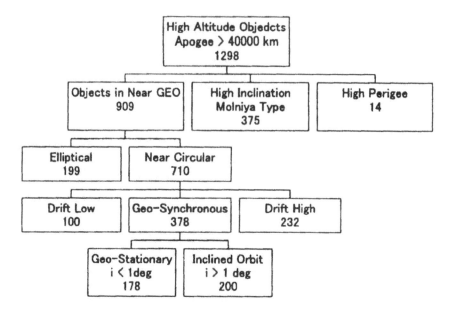

Fig. 5.3. Near GEO objects: groups and numbers.

sub-divided into three groups according to their orbital periods. Less than 2 minute difference from nominal geo-stationary period of 1436.07 minutes is chosen for selection of "Geo-Synchronous" objects from other objects in drift orbits, "Drift Low" for shorter period and "Drift High" for longer period. "Geo-stationary" objects are a special kind of "geo-synchronous", where orbital inclination is kept small by periodic orbit control maneuvers. All of 178 "geo-stationary" objects are considered operational satellites. 200 objects in "Inclined Orbit" with larger than 1 degree inclination are most probably expired satellites that lost control at the geo-stationary orbit, either accidentally or intentionally. On the other hand, those that were commanded to re-orbit at the end of life now belong to "Drift High" group.

Until 1992, no breakup in geostationary altitude were publicly known. In February of this year, Russian scientists revealed an astonishing evidence of spacecraft explosion. In-orbit explosion of EKRAN communication satellite was photographed by a ground telescope. It was explained that this phenomenon was recorded more than ten years before. At about the same time as the EKRAN photograph release, US Maui surveillance station observed a fragmentation event of a 24-year-old spacecraft drifting at near geostationary altitude. Up to now, three breakups have been reported including above mentioned events. It should be recognized that observation of these events are extremely difficult to be performed by ground based facilities because of a great distance to geo-synchronous altitude in comparison with detection of similar events in low earth orbit. More breakups might have taken place without being noticed. There is not enough data on how many fragments were created by those breakups, but it is quite certain that there are a number of fragments already orbiting in this altitude without being detected.

5.1.2. Collision Hazard

Historically, communication system operators, who compose the largest user group of this orbit, first became aware of physical interaction of satellites in geostationary orbit. After extensive discussions, in June 1992, CCIR Working Party 4A adopted a draft recommendation (CCIR, 1992) that satellites in GEO be re-orbited into higher orbit at the end of life, to lessen congestion in geo-synchronous altitude. This became effective in the following year. Even before the adoption of the recommendation, many operators did re-orbited their satellites, and now, the necessity of this maneuver is recognized universally.

Communication satellites are sensitive to other satellites that come close by because of radio interference. Therefore, stationary orbit slots are coordinated by ITU. An orbital slot is occupied by the same series of satellite after an aged satellite retires. Due to the nature of these satellites that are "stationary" at a certain longitude above the equator, there comes a consideration of collision hazards between the new operational and the expired satellites. The concept of re-orbit maneuver is derived quite naturally. Actual evaluation of collision hazards was considered by Perek (1989), Chobotov *et al.* (1989), Yasaka (1991) and others. The rate of a satellite being hit by another, estimated in late 1980 time frame, was in the order of 10^{-7}–10^{-8} /year/m^2, which generally resulted in a much lower value than failure rate of communication equipment or other satellite functions. It is difficult to blame a number of satellite operators that had been reluctant to accept the re-orbit policy at their satellite's end of life, when this maneuver is evaluated as equivalent to a few months shortening of operational life. However, accumulation of increasing number of objects suggested that the hazard would soon reach a level where even a short-term interest of operating group might be jeopardized if the re-orbit maneuver were not universally conducted.

Collision rate of an object with other objects is proportional to density, relative velocity and cross sectional area. Density shows a sharp peak in a narrow ring, whose center is the nominal geostationary orbit of 42165 km radius, thickness is 50 km in altitude and a few hundred km in latitude directions. In the latitude direction however, a rather high density region extends further to a few thousand km, where objects in inclined orbits are located. The velocity with reference to the Earth is very small in case of controlled objects. Orbit correction maneuvers are periodically conducted, imparting delta velocities of around 10 m/s and 0.2 m/s in north-south and east-west directions, respectively. These velocities represent the magnitude of velocity differences among controlled satellites. On the other hand, objects in inclined orbits have a much larger velocity component in north-south latitude directions, in proportion to the inclination. These objects penetrate the equatorial plane, through the dense ring, at 800 km/s maximum, twice a day, as it is seen in Fig. 5.2 trace pattern. Considering these discussions on density and velocity, collisions occur most likely within the dense ring of operating satellite by fast penetrating objects in inclined synchronous orbit.

A collision occurrence number C_T in a prescribed time period is sometimes very important. This is an integration of the collision rate by time. Simply expressing the collision rate as a product of velocity V, density ρ and cross sectional area S,

$$C_T = V\rho S \cdot \Delta t \cdot n \tag{5.1}$$

where Δt is the time for the object to pass through the dense ring, and n is the number of traverse the object makes during the time period considered. Assume that operational satellites are uniformly distributed in a ring of radius R, whose cross section is rectangular with the length L in latitude and H in radial directions. Since the object travels the length L at a velocity V, $\Delta t = L/V$. The density is obtained from the total number of operational satellites N divided by the volume of the ring. C_T is evaluated over a revolution ($n = 2$) and,

$$C_T = 2\rho S \cdot L = \frac{2N \cdot S}{2\pi R \cdot H}. \tag{5.2}$$

The velocity is not actually important as long as the object penetrates the ring. The number of operating satellites and its dispersion in radius direction is influential to the collision occurrence. Considering a normalized collision rate which is defined here as a collision probability of an operational satellite ($N = 1$) with a penetrating object during operation of one year, while the collision cross section is unity ($S = 1$ m^2), one obtains $5 \cdot 10^{-11}$ collisions per year per square meter of cross section, by substituting 50 km to H.

An object in a slightly eccentric orbit also intersects with the dense GEO ring at an appreciable velocity, which is proportional to the eccentricity. Following the similar reasoning as above, and after a considerable manipulation in passage time and relative velocity as described in the reference (Yasaka, 1993), collision rate during a revolution is obtained in a form:

$$C_T = 4\rho S \cdot H = \frac{4N \cdot S}{2\pi R \cdot L}. \tag{5.3}$$

In this case, equatorial in-plane motion is assumed, while the former case of inclined orbit deals with out-of-plane motion. Influential parameters in both cases are thickness of the ring, but the role of L and H are inverted.

Collision rates among orbiting objects are calculated by using actual object catalogue data in 1997, as shown in Tables 5.1 and 5.2 (Yasaka, 1998). There were 632 objects considered grouped to geo-stationary (GST), inclined geo-synchronous (GSY), drift-high (DRH) and drift-low (DRL) in a similar manner as in Fig. 5.3. The collision rate between operating satellites in geo-stationary orbit and uncontrolled objects in inclined orbits is the most significant, and its normalized value agrees well with a simple estimate based on the equation (5.2).

Table 5.1. Collision rates among 632 objects in different classes (10^{-6} collisions/year/m^2)

	GST	GSY	DRH	DRL
GST	0.786	–	–	–
GSY	1.329	0.620	–	–
DRH	0.086	0.077	0.049	–
DRL	0.059	0.050	0.002	0.001

Table 5.2. Normalized collision rates between two objects (10^{-11} collision/year/m^2)

	GST	GSY	DRH	DRL
GST	2.86	–	–	–
GSY	4.17	1.68	–	–
DRH	0.25	0.20	0.12	–
DRL	0.51	0.38	0.01	0.02

5.2. Breakup Model

Various mathematical models have been constructed to describe breakup processes in space. They are mostly based on empirical data, either of ground observation of in-orbit breakups, or on hyper-velocity impact laboratory tests. These models can be applied to hyper-velocity impacts usually observed in low earth orbits. Collisions in GEO, on the other hand, occur at far slower velocities because of lower orbital velocity and low incident angles of below 15 degrees. Therefore, there is no way to guarantee the validity of those established models when they are applied to GEO collisions. The model described here aims at estimating mass distribution and associated velocities of fragments based on basic physical principles, conservation of momentum and energy, without relying too much on empirical data.

5.2.1. Momentum and Energy Relations

A target object and a projectile whose masses are m_t and m_p, and orbital velocities are V_t and V_p, respectively, are assumed to collide. At the collision, they breakup into n fragments whose masses and initial velocities are m_i and V_i ($i = 1 \ldots n$). Momentum conservation is expressed as:

$$m_t V_t + m_p V_p = \sum m_i V_i \tag{5.4}$$

Denoting that

$$V_0 = \frac{m_t V_t + m_p V_p}{m_t + m_p}$$

and

$$V_i = V_0 - v_i \,,$$

equation (5.4) now becomes

$$\sum m_i v_i = 0. \tag{5.5}$$

Defining relative velocity at collision as $V_r = V_p - V_t$, kinetic energy before the collision is expressed as:

$$E_0 = \frac{1}{2}m_t V_t^2 + \frac{1}{2}m_p V_p^2 = \frac{1}{2}(m_t + m_p)V_0^2 + \frac{1}{2}\frac{m_t m_p}{m_t + m_p}V_r^2.$$ (5.6)

Kinetic energy after the breakup is

$$E = \frac{1}{2}\sum m_i V_i^2 = \frac{1}{2}(m_t + m_p)V_0^2 + \frac{1}{2}\sum m_i v_i^2.$$ (5.7)

The difference $E_b = E_0 - E$ represents the breakup energy which is consumed in the process of fragmenting objects into n pieces. The minimum E_b is zero when all energy is transmitted to fragments, and the maximum is the second term of right hand side of equation (5.6) or $V_r^2 m_t m_p/2(m_t + m_p)$, in which case all v_i are zeros. In an actual breakup, E_b takes a value which is somewhere between the minimum and the maximum value. Let us introduce a parameter λ as

$$\lambda = \frac{E_b}{(E_b)_{\max}} = 1 - \frac{\sum mv_i^2}{\left(\dfrac{m_t m_p}{m_t + m_p}\right)V_r^2}.$$ (5.8)

The parameter λ represents a ratio of the energy of breakup to the theoretical maximum, and its value lies between 0 and 1. Rewriting the equation, one obtains a simplified relation

$$\sum m_i v_i^2 = \frac{m_t m_p}{m_t + m_p}V_r^2(1-\lambda).$$ (5.9)

Equations (5.5) and (5.9) are the relations that must be always satisfied by a set of m_i and v_i.

It is interesting to note that in case of another kind of breakup in which internal energy stored in an orbiting object provides the fragmentation energy, a similar formulation in place of equation (5.9) can be derived.

$$\sum m_i v_i^2 = 2E_c(1-\lambda)$$ (5.10)

where E_c is the energy released at the breakup, and $\lambda = E_b/E_c$.

5.2.2. Velocity Distribution

Velocities of fragments after the breakup can widely distribute, and in many cases, they can be represented by random variables. In order to ensure the momentum and energy conservation laws, the randomly generated velocity vectors are adjusted in a following way.

(a) Randomly generated velocity vectors $\bar{v}_i$ are related to a set of possible velocity vectors v_i through a constant bias v_b and a scale factor q.

$$\bar{v}_i = qv_i + v_b \tag{5.11}$$

(b) The bias v_b is determined by using equation (5.5) as:

$$v_b = \frac{\sum m_i \bar{v}_i}{\sum m_i}. \tag{5.12}$$

(c) The scale factor q can then be evaluated either by equation (5.9) or (5.10).

$$q^2 = \frac{\sum m_i (\bar{v}_i - v_b)^2}{\left(\dfrac{m_t m_p}{m_t + m_p}\right) V_r^2 (1 - \lambda)} \tag{5.13 a}$$

or

$$q^2 = \frac{\sum m_i (\bar{v}_i - v_b)^2}{2E_c (1 - \lambda)} \tag{5.13 b}$$

Vectors v_i obtained using equation (5.11) satisfy minimum requirements of momentum and energy conservation which must be met by all breakups of any type. As long as velocities are not explicit functions of mass, breakup events can be treated in this manner. Characteristics of a breakup can be largely dependent on the mass distribution and energy transfer rate.

5.2.3. Mass Distribution

In the forgoing discussion, mass distribution was separately given. The simplest would be to assume fragments of identical mass. However, experience tells us that smaller fragments are larger in number at any breakup. A simple and good estimation of the mass distribution, in case of a hyper-velocity impact laboratory tests, is given by a power law, first introduced by Su and Kessler (1985), in a form

$$CN = a(m_f/m_e)^{-b}$$

where CN is the cumulative number of fragments with mass m_f or larger and m_e is the total ejecta mass. Constants a and b depend appreciably on types of experiments, including target and projectile materials. Application of this hyper-velocity empirical relation on mass distribution into the present procedure generally results in a typical collision induced fragments dispersion pattern when evaluated in the Gabbard Diagram. Also, velocity envelope of fragments against mass is consistent with experimental characteristics, where smaller fragments generally get larger velocity increments.

5.2.4. Verification of Assumptions (Yasaka, 1996, 1997, 1988)

The present breakup model is developed so that it may be applicable to collision events in GEO, where the typical collision velocity is an order of magnitude smaller than those in LEO. A specially designed apparatus is used to validate assumptions on impact fragmentation at simulated velocity range of GEO collisions. This is a single stage air-gun, which accelerates a projectile up to a few hundreds meters per second. A CCD camera detects fragments images at separate time intervals under multi-flashed strobe light. The impact test chamber is a closed box, and ejected fragments are very carefully collected for latter examination. The projectile is a 9 mm steel sphere and the target is a CFRP honeycomb sandwich panel. Several tests were conducted at projectile velocities between 110 m/s and 160 m/s. In all cases, the projectile just penetrated the target panel, leaving a hole slightly larger than the projectile diameter. In an attempt to evaluate fragments mass distribution, all fragments in the test chamber were recovered including very tiny dust size pieces. The total mass of recovered particles was about 50 mg, which was identical to the lost mass in the target panel. A very careful evaluation of fragments resulted in the mass distribution that could be well fitted to a power law as shown in Fig. 5.4. In case

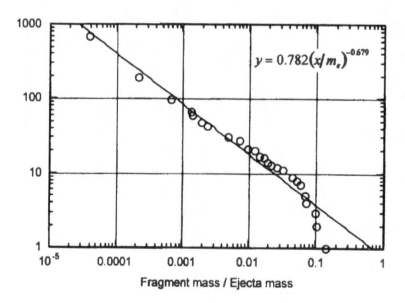

$$y = 0.782(x/m_e)^{-0.679}$$

Fragment mass / Ejecta mass

Fig. 5.4. Mass distribution of fragments ($V_p = 150$ m/s).

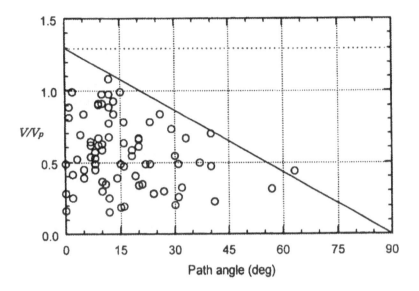

Fig. 5.5. Fragment velocity ($V_p = 150$ m/s).

of hyper-velocity experiments, there has been an argument that the cumulative number curve levels off for small masses, which led to a suggesting parabolic law instead of the power law. The present low velocity impact test results tell us that the power law is effective down to the smallest mass range where the measurements are still sufficiently accurate.

Fragment velocities were measured individually by evaluating images taken by the CCD camera under two strobe flash lights at less than a milli-second apart. From a plot of fragment velocities against the path angle measured from the initial projectile's path, the largest velocity was a little bit larger than the projectile's initial velocity, as shown in Fig. 5.5. It should be noted that since these tests do not satisfy the momentum conservation, the velocity distribution can not be obtained by the theory stated in 5.2.2. However, the qualitative pattern of the velocity profiles obtained from the theory agrees well with the experiments.

5.2.5. Debris Cloud Evolution

By applying the velocity profile in 5.2.2 to an orbit propagator, one obtains a debris cloud which changes its shape as time goes on. The cloud starts to elongate arms into both east and west directions from the point of breakup. In a matter of weeks, both arms make a complete revolution around the earth, but the arm tips do not touch each other, because orbital radius of east bound object is smaller than that of west bound objects. After several revolutions of the arm tips, the cloud forms a spiral shape, as shown in Fig. 5.6. Each time the arm tips make half a revolution, a spiral fringe is added accompanied by reduction

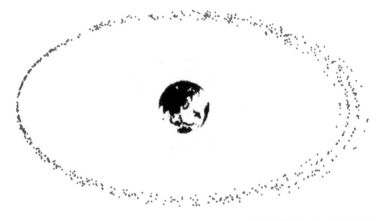

1996/08/15 16:00:58 JST
Space Systems Dynamics Lab., Kyushu Univ.

Fig. 5.6. Fragment cloud, 24 days after breakup collision of a geostationary satellite and an object 2 degrees inclination.

of fringe separation distance and blurring of fringe pattern, until the cloud becomes a single disk. Orbit perturbation by solar radiation pressure accelerates the fringe blurring.

Collision rate of an operating satellite with breakup fragments shows periodic peaks, when the cloud arm passes by the satellite. As the spiral pattern blurs with time, the rate levels out to a value, which is smaller for the higher altitudes of breakup.

5.3. Objects Number Evolution Modeling

The absolute objectives of GEO debris modeling are to find out how the orbital debris environment will change in time, what parameter affects the growth rate of debris number and what could be the most effective method to maintain the environment under control. There are precise and comprehensive debris models already established, including MASTER (Klinkrad *et al.*, 1997), EVOLVE (Johnson *et al.*, 1997) and CHAIN (Eichler and Rex, 1989). They are all based on empirical breakup models and their performances have been tested and verified by comparison mainly with the present LEO debris environment. Capability of precise simulation of the present debris environment will enable forecasting future situations as an extension. The applications of these models to GEO environment could be also possible. But since the data on small size debris for the models to be verified against is not available in GEO, the results provided by the above models should be looked upon differently from those of LEO. The method described in this section does not intend to precisely simulate the present debris environment, but the method is aimed at identifying influential parameters from the point of view of hazard to operational geostationary satellites, using simple mathematical procedures.

The only sources of man-made objects in GEO altitudes are launches routinely conducted since 1963. Those objects injected into this orbit have a very long lifetime and they remain in the vicinity of nominal geo-stationary orbit under natural perturbing forces and even after intentional re-orbiting maneuvers. Therefore, the mass to be considered is added at each launch and, unlike the case in LEO, the mass is never removed. However, in considering hazards to other satellites, the number of objects is of most concern rather than the total mass, because even small mass could be fatal to operational satellites in collision. In this chapter, any event that increases the number of objects is considered as a source. Sources considered here are explosions, collisions and launches. Disintegration of aged satellites can also be a serious source of objects increase. However, this will not be included because we still do not have enough knowledge of these phenomena. The desintegration of aged satellites should be treated in future.

5.3.1. Evolution Model

The program GEO-Evolve has a purpose to estimate objects number in GEO and its vicinity by taking into consideration launches, collisions and explosions. Objects are grouped according to their orbit properties, mainly orbital period/altitude and inclination. For the sake of simplicity and without any loss of generality, all object are assumed to have small eccentricities. The variable $X_1(t)$ is the number of operational satellites as a function of time. They are always controlled to stay within a narrow geostationary ring. $X_2(t)$ is the number of objects in inclined orbit, but still maintaining earth revolution period. They are normally aged satellites left in the original altitude without re-orbiting maneuver to raise altitude. $X_3(t)$ represents objects in drift orbit either higher or lower than the geostationary altitude and they do not intersect with the area where operational satellites are located. $X_4(t)$ and $X_5(t)$ are numbers of debris which were created by breakups of other objects. Normally they are not geo-synchronous any more, but the former ones cross the geostationary altitude and the latter do not. The numbers that we are mostly interested in are $X_2(t) + X_4(t)$, which show the level of possible hazard to operational satellites.

Assuming that life spans of all operational satellites are the same, L years, and taking one year interval for incremental time step, the increments of these variables are expressed in simple equations as:

$$
\begin{aligned}
\Delta X_1(t) &= R(t) - R(t-L) - Y_1(t) \\
\Delta X_2(t) &= (1-p)\,R(t-L) - Y_2(t) \\
\Delta X_3(t) &= pR(t-L) - Y_3(t) \\
\Delta X_4(t) &= n_1 Y_1(t) + n_2 Y_2(t) + \beta n_3 Y_3(t) \\
\Delta X_5(t) &= (1 - \beta)n_3 Y_3(t)
\end{aligned}
\qquad (5.14)
$$

where $R(t)$ is the annual launch number in the year t, p is the re-orbiting rate or the number fraction of objects which are commanded to re-orbit among all the end-of-life satellites. $Y_i(t)$, $i = 1, 2, 3$ are numbers of breakups among objects in the i-th group within a year time, creating each time n_i fragments. All fragments created in breakups of objects belonging to groups 1 or 2 cross the nominal GEO altitude, where the parent objects were originally located. Consequently, these fragments are added to X_4. On the other hand, fragments released by objects belonging to the group 3 do not reach GEO altitude and are added to X_5 if the energy imparted at the breakup is small enough. But otherwise, some fragments do reach GEO and their number should be added to X_4. The ratio is designated by β, which is a migration ratio of fragments from the harmless altitude to a hazardous orbit.

The breakup number is a sum of collisions and explosions.

$$Y_i(t) = \sum_{j=1}^{5} C_{ij}(t) + \gamma X_i(t)$$

$$C_{ij}(t) = \alpha_{ij} X_i(t) X_j(t)(A_i + A_j)$$

$C_{ij}(t)$ is the collision number between sets of objects in groups i and j within a year time, which is proportional to the number of objects in each group and cross sectional areas of the objects. The proportionality constant α_{ij} is the normalized collision rate which is defined as an average collision rate of one object in the group i and another object in the group j, each having a unit area of cross section. Explosions number is simply a product of the explosion rate γ and the number of objects.

5.3.2. Parameters and Simulation Results

Equations (5.14) contain several parameters. Some of them reflect trends of future space activity and technology efforts, and the others depend on physical characteristics of breakups. The list of parameters includes the launch rate $R(t)$, re-orbit ratio p, life L and explosion rate γ. The purpose of this modeling effort is to identify the influential ones among these parameters, which have a determining influence on the objects number growth in the geo-stationary orbit. Other breakup related parameters, mainly the normalized collision rate α_{ij}, migration rate β and fragment numbers n_i, are of statistical nature, which should be determined by using collision estimation analyses and appropriate breakup models.

Selection of the parameters values and the resulting number growth trend that first appeared in the literature by Yasaka (1993, 1994) resulted in the following list of values.

(1) Launch rate and satellite life: Starting from 25 launches per year in mid 1990's, the number of launches is increasing by one each year. After reaching the maximum of 100, the rate is kept constant. All satellites are assumed to function for 10 years.
(2) Collision rate: Any two objects in various groups are set to have normalized collision rate similar to those shown in Table 5.2. The cross sectional area of objects are assumed to be 100 m^2 for intact objects in groups 1 through 3, and 1 m^2 for fragments in groups 4 and 5.

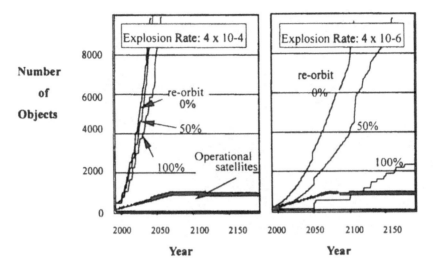

Fig. 5.7. Number growth of hazardous objects in GEO.

(3) Explosion rate: 4×10^{-4} per year per object is set to be the reference value. This value was derived from the known explosions in GEO and the total assembly of objects in this region. In the year 1993, there were 505 objects whose average time spent in orbit after launch was 10 years. Two explosions were known at that time. The explosion rate is obtained by $2/(505 \times 10)$.

(4) Breakup fragments: At any breakup, 300 fragments are assumed to be released. The migration ratio is assumed simply to be 0.5.

(5) Initial conditions: $X_i(0)$ ($i = 1\cdots5$) are set to be (120, 149, 246, 0, 0), reflecting catalogued objects number in 1993.

(6) Re-orbiting ratio: In the earlier time of GEO utilization, all satellites were left in the original orbit, or $p = 0$. Now it is requested that all satellites should be orbit raised, or $p = 1$. Various values are assumed and kept constant over time, to investigate contribution of the re-orbiting ratio.

The simulated results are shown in Fig. 5.7. Thin lines show numbers of total objects that cross GEO altitude, which can be hazardous to operational satellites. Explosion rate is chosen as the main parameter and re-orbit ratio is a secondary parameter. As long as the explosion rate remains at the reference value, the number of hazardous objects increases uncontrollably within 50 years. When the explosion rate is chosen two orders of magnitude less, the re-orbit ratio becomes influential. Different initial conditions also give very similar results (for instance, Yasaka, 1988).

In order to quantitatively express the level of hazards to operational satellites, the estimated number of operational satellites lost during their life span is shown in Fig. 5.8. Nearly 100 satellites could suffer damage within 200 years. One fourth of them are collided by debris particles or other aged objects. Three fourths are expected to explode themselves.

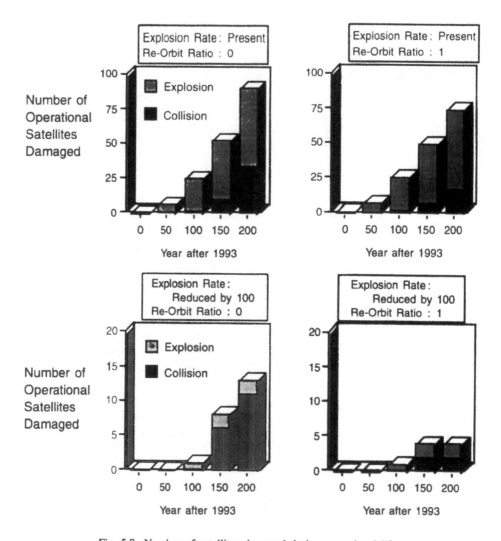

Fig. 5.8. Number of satellites damaged during operational life.

If the explosion rate is two orders of magnitude less, the self-destruction reduces at the same ratio, and collisions now become relatively significant. However, absolute number of collisions does essentially reduce, reflecting the decrease of hazardous objects.

Some questions still remain, as to selection of parameters values: the migration rate β and fragments numbers n_i. Using the breakup model suggested in 5.2, under a specified energy level and breakup altitude, the migration ratio can be estimated after applying the orbit propagation tool to objects characterized by initial breakup velocities. Recently, Kyushu University formulated an empirical relationship to convert combination of energy level and breakup altitude into the migration ratio. The minimum fragment mass is readily obtained from the power law relationship in 5.2.3, once the total fragments mass is assumed. The number of fragments is then determined considering the critical mass that would affect the satellite. The above example corresponds to the minimum fragment mass of 0.15 kg in case of a total breakup of 1000 kg object. However, selection of different values for these parameters does not alter the overall trend appreciably. In all the cases, the explosion rate remains the most influential parameter, and the debris number growth reaches uncontrollable rates sooner or later. Collisions do not play the main role in the GEO population growth, but rather in-orbit explosions.

5.4. Necessary Preventive Measures

5.4.1. Graveyard Orbit

Geostationary communication satellite operators became aware of the necessity of elimination of expired satellites from the useful orbit and came to an idea of graveyard orbit. A very long discussion was held on the effectiveness and affordability of end-of-life disposal of spacecrafts from GEO, before this policy obtained the world wide consensus. The purpose of this idea was to secure safety of satellites by avoiding satellite to satellite collisions. If one admits that aged satellites finally turn into small pieces of objects during a long time exposure to space environment, by whatever the cause, the concept of graveyard orbit is found not effective to avoid eventual contamination of GEO. In a middle term future or at least in the next hundred years or so, however, the concept works well. Re-orbited objects scatter in a large volume of space and their mutual collision rate is low. As long as objects remain intact or large enough, being not susceptive to natural perturbations, they are confined within the harmless region. The graveyard option effectively eliminates masses from GEO for the time frame we are currently facing.

5.4.2. Explosions Control

Explosions in geo-synchronous altitude brought up a new kind of hazard that the graveyard concept cannot handle properly. An explosion creates hundreds of fragments which are large enough to cause failure of operational satellites in collisions. Generally these fragments form a large debris cloud whose thickness in radial directions is more than a few thousand km. Therefore, explosions within a few thousand km from geo-synchronous altitude create hazardous fragments to all objects in the vicinity of GEO. If the present explosion trend continues for all objects in the future, the number of objects will soon show a runaway. The immediate measure to maintain safety is to reduce explosion rate of all objects to be launched into geo-synchronous orbit by about two orders of magnitude.

Technical solutions for explosion prevention are not different from actual measures being proposed and executed for low Earth orbiting objects. The primary philosophy is exhaustion of energy source from any object prior to control loss. In addition, satellite design should be examined so that the explosion possibility was minimized. In low Earth orbits, depletion of residual propellant from upper stages boosters after completion of mission was proved to be sufficiently effective to avoid explosions. In fact, all boosters which depleted residual propellant on command have not experienced explosions in orbit. The similar maneuver alone would decrease the explosion rate appreciably in GEO as well. In contrast to LEO objects, since orbital life in GEO altitude is considerably longer, it is desirable to vent all closed containers including pressure vessels and eliminate all explosive devices which might be left unused until the end of life. Presently executed re-orbiting maneuver contributes to energy depletion as well, since residual propellant and high pressure gases are exhausted during the maneuver.

5.4.3. Other Debris Sources

Collisions, explosions and launches were considered as the sources of objects accumulation in this chapter. However, existence of other sources must be also considered. In the low Earth orbits, leakage of liquid metal from aged spacecraft vessels is suspected to be a new source of debris. There are possibilities of completely new kind of debris source identifications, but there also exist a logically possible source that has not been considered properly. The latter is the desintegration of spacecrafts due to material degradation during a long time exposure to space environment.

Generally adopted procedure of endurance verification of spacecraft materials and components are to examine their performances during the spacecraft operational life span. Over-testing in excess of useful life period is not recommended because it would add unjustified cost to the spacecraft development. Consequently, we are completely in the lack of knowledge on the fate of spacecrafts after the life span, whose maximum is a little bit more than ten years. However, it is easy to imagine solar cells coming free from the solar cell substrate, exterior thermal control materials detached form the spacecrafts walls, or fibers getting loose in fiber reinforced plastic structural components, while drifting in space for much shorter than a hundred years. Most of organic materials now used to build satellites loose their strength under repeated thermal cycling and radiation bombardment. Non-organic materials would also behave similarly, after a considerably long exposure. The logical consequence is that all spacecrafts will gently come apart and turn into many pieces of debris. And it is not known absolutely when it could happen.

5.4.4. Necessary Measures

Continuation and complete execution of re-orbiting maneuvers of satellites at the end of their life are important. Since re-entry into the Earth's atmosphere or escape from the Earth's orbit are practically unfeasible options for geostationary satellites due to large delta velocity requirement, the re-orbiting into the graveyard is the only feasible choice to eliminate unnecessary masses from the useful altitude region. Explosions pose much larger threat to the orbit contamination by small debris fragments. Immediate actions should be taken by all parties concerned to lessen the explosions possibility. Most of technical measures are already known and their effectiveness has been verified by applications to LEO objects. It should be recognized that we have already accumulated a large number of potentially hazardous objects. Immediate prevention of further accumulation will allow us time to cope with future hazards, but delays result in a more costly future operations to maintain the safe environment, if unrecoverable contamination could be avoided.

Contamination by gradual desintegration of spacecrafts seems unavoidable. It is not the author's intention to require replacement of the present spacecraft materials by longer durable materials. Investigations in this area would be quite beneficial, but what is more important at present is to undertake investigations of the actual behavior of aged satellites staying now in orbit. The most effective measure could be identified through the knowledge of the present situation and the knowledge of its trends. Actually, no agencies or firms will become enthusiastic to attack this problem until evidence of unrecoverable contamination is identified. A small probe with a camera and rendezvous capability to orbiting objects will provide us with necessary information to get started. By getting images of several sample objects representing various exposure time period in space, we will gather enough knowledge to simulate the environment in years to come.

Active intervention to orbiting objects will eventually become necessary. This applies not limited to GEO objects, but generally to all objects in high orbits without sufficient atmospheric drag to reduce orbital life. Especially in GEO, where re-orbiting is essential for environment protection, expired satellites left in the geo-synchronous altitude must be managed somehow. Some satellites might find failure to respond to re-orbiting command in future. They must be removed likewise. A service vehicle with a grapple fixture can effectively re-locate these uncontrolled objects (Yasaka, 1993). The graveyard orbit is a short term candidate for the place of re-location. A storage depot is a long term candidate for collective disposal out of the Earth orbits. The most desirable future is to re-cycle the spacecraft materials stored in the depot for construction of other permanent facility in the orbit. Depending on the desintegration characteristics of objects described in 5.4.3, the service vehicle would manage aged objects before they turn into small debris, in a manner which could be identified to be most effective after the investigations by simpler probes or service vehicles.

Removal of small debris does not seem to be a good option. Capturing it will be too costly. There might be a novel device to reduce their velocity, to remove them into lower orbit, but there is no effective one known to us today. Capturing before they turn into small pieces would be the most cost effective and feasible. However, a legal issue must be pointed out as an important step to implementation of the service vehicle capture strategy. It is clear that an active satellite must not be captured and handled by anybody without a legal consensus of the operator and the owner. An object that ceased to function is generally termed as a debris, but legal status of debris is not yet made clear. It is most likely, at present, to assume that these capture operations are done under agreement of two parties, one who legally owns the object and the other who conducts the operation.

CHAPTER 6

AREA/MASS AND MASS DISTRIBUTIONS OF ORBITAL DEBRIS

P.D. Anz-Meador and A.E. Potter

Area-to-mass ratios for orbital debris tracked by the U.S. Space Command were calculated from observed changes in semimajor axis due to atmospheric drag. The area-to-masses observed for the orbital debris were similar to those found for debris from laboratory breakups, and suggest that much of the debris is composed of crumpled thin plates or of insulation material with low effective density. Areas for the debris objects were derived from radar cross-section (RCS) data. Object masses were calculated from the ratio of the RCS-derived area to the area-to-mass ratio. Analysis of the distributions of fragment masses from the breakups suggests that in many cases, only a portion of the initial object breaks up into small fragments.

6.1. Background

Most of the debris in EARTH orbit has been produced by explosions or collisions. In order to estimate the population of debris with sizes below that, which can be tracked by ground-based radars and telescopes, it is necessary to model the number and size distribution of debris from the breakups. The classic work on modeling of breakups was reported by Bess in 1975 (Bess, Dale, 1975), who analyzed the fragment distribution resulting from laboratory explosions and hypervelocity collisions. His work is the basis of the breakup models used by environment models such as EVOLVE (Reynolds, 1990, 1991), CHAIN (Eichler, Rex, 1992), and many other current debris models. However, during the past ten years, data from many on-orbit breakups has accumulated. It is possible to learn something about the true nature of these breakups from analysis of these data. Tools for this analysis have

been developed by Badhwar and Anz-Meador (1990, 1992), using orbital decay to calculate area-to-mass ratios, and radar cross-sections to calculate areas. From these data,
both mass and area of the fragments can be derived. Their procedure for calculating the
area-to-mass ratio of debris fragments from decay of their orbits over time is documented
in reference (Badhwar, Anz-Meador, 1992). Their procedure for calculating areas from
radar cross-sections (RCS) is documented in reference (Badhwar, Anz-Meador, 1990). For
the latter, they developed a relationship between area and RCS, using RCS values from
U.S. Space Command (USSPACECOM) measurements and the areas of intact satellites
of known dimensions (Badhwar, Anz-Meador, 1990). They tested the overall procedure
for calculating areas and masses by using it to compute the masses of all the fragments
from several breakups at high altitudes where most of the debris was still in orbit. The
sum total of all the masses calculated in this way was in good agreement with the initial
mass of the object that had broken up, thus supporting the overall correctness of the
method. We used this technique to calculate the area-to-mass ratios and the masses of
individual debris fragments for a number of breakups. The results can be compared with
theoretical predictions from breakup models to test their validity for on-orbit breakups.

6.2. A Case in Point: The Fragmentation of Cosmos 1484

On 18 October 1993, the derelict *Cosmos 1484* Earth resources spacecraft fragmented.
A precursor of the current *Resurs*-class spacecraft, this event was unusual in that the
spacecraft had been on orbit for ten years and was one of the rare Soviet payloads in Sun
synchronous orbit. Both of these observations led to an early supposition that the event
may have been caused by a collision with a tracked or unseen object. Analysis quickly
revealed that no tracked objects were in the vicinity of the payload at the event time and
that *Cosmos 1484* had indeed undergone a sudden change in its mean motion and thus
was the progenitor of approximately 79 debris objects (Nauer, 1993), of which 48 were
eventually cataloged by the USSPACOM. However, ground-based observations of *Cosmos
1484* by the German FGAN radar indicated that the spacecraft was essentially intact.

Sets of orbital elements ("elsets") of the debris fragments were provided to NASA by
the USSPACECOM. Orbital decay data from these elsets were analyzed to yield values
for the area-to-mass ratio, using techniques described in (Anz-Meador, Rast and Potter,
1993). In order to determine masses from these ratios, the areas were required. These were
determined from radar cross-sections (RCS) as noted above, but with further refinements
as outlined below.

We have collected over 50 USSPACECOM RCS catalogs spanning 1977-1999. From
these, we have derived a statistical description (Anz-Meador, Henize and Kessler, 1994)
of the cataloged objects expressed by a median, mean, standard deviation, and higher order
statistics. Approximately 85% of these objects exhibit normality (David, Hartley and
Pearson, 1954). However, the RCS medians are utilized to minimize the scatter caused
by statistical outliers and are converted to characteristic sizes using the NASA Size
Estimation model (Bohannon, 1992). This model has been developed from laboratory
measurements of radar cross-section for a wide variety of fragments from ground-based

Kosmos 1484 Breakup Debris

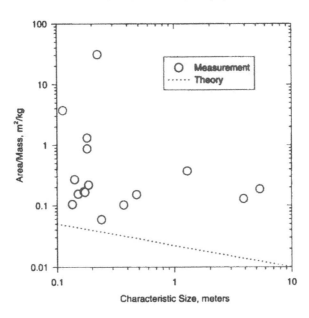

Fig. 6.1. Area-to-mass ratios for kosmos 1484 fragments.

explosion tests. With the assumption that the area A may be represented by a circle of diameter d, we solve for the mass: $m = \pi d^2/4\xi$, where ξ is the area-to-mass ratio.

A radar cross section catalog was obtained that contained RCSs for the majority of these objects. After conversion of RCS values to characteristic area as described above, and dividing by the area-to-mass ratio, we obtained the mass of each object.

The area-to-mass data are plotted against size in Fig. 6.1, where they are compared with a solid line that corresponds to the area-to-mass ratio assumed in the EVOLVE 3.0 (1998 current) orbital debris environment model (Reynolds, 1990, 1991). Area-to-mass ratios are an order of magnitude larger than expected from the model. The debris objects have much lower effective densities than would be expected if they were fragments of conventional spacecraft structures. The total mass of all the debris objects was only 16.7 kg. It was evident that only a small part of the satellite exploded, and that the fragments were of some light-weight material, such as sheets of thermal insulation material.

The Russian manufacturers of the spacecraft suggested that the breakup could have resulted from explosion of an ammonia tank attached to the side of the satellite. We analyzed the energy available from the pressurized ammonia tank, and what would happen should the tank rupture. Our results supported their suggestion. We concluded that gas pressure in the tank could have exceeded the design pressure limit due to solar heating that resulted from changes in spacecraft attitude and deterioration of the thermal control coatings. Some of the energy of gas expansion was spent rupturing the tank, and the remainder had gone into an impulsive thrust acting on the spacecraft hulk and the tank debris. The light-weight tank debris was accelerated to considerable velocities. The massive spacecraft hulk was affected only slightly. Similar events are surmised (Grissom, Guy and Nauer, 1994) to have occurred in the *Ekran 2*, *Nimbus 7* rocket body, *NOAA 8* rocket body, and other fragmentations, although these fragmentations have not been analyzed in the manner just described.

This event, so different in nature from that usually assumed for the fragmentation of a satellite body, prompted a systematic investigation of a representative sample of breakups, in order to define some of the real circumstances of on-orbit breakups, as opposed to the supposed circumstances of the breakups. We reported on that initial study in 1994; this work updates that study.

6.3. Area-to-Mass Ratio from Orbital Decay

The area-to-mass ratio ξ was calculated for each object in a debris cloud from the secular change of semimajor axis as the orbit decayed. Semimajor axis, rather than apogee and perigee altitudes, was used in an effort to minimize eccentricity-driven orbital effects due to higher-order gravitational Geopotential terms (J3 and higher) and Solar radiation pressure. Calculations were done as reported previously (Anz-Meador, Rast and Potter, 1993), using the Mueller atmospheric decay model to model the orbit decay with time for any given area-to-mass ratio, and assuming a Jacchia exospheric temperature profile that was modified to reflect saturation at higher values of the 10.7 cm solar flux. We performed this analysis using a longer series of orbital decay data than previously available (Anz-Meador, Rast and Potter, 1993). Brent's inverse parabolic interpolation method (Brent, 1973) and the object's time-dependent semimajor axis (a) was used to estimate the area/mass ratio between adjacent elset observations. A noise-floor filter in area/mass-da/dt (rate of change in semimajor axis with respect to time) space was used to remove outliers resulting from bad observations, cross tagging, *etc*. The resulting ensemble of observations was then processed to yield descriptive statistics of median, mean, standard deviation, skewness, and kurtosis. While the majority of ensembles displayed a normal or near-normal frequency distribution in area/mass, the median was chosen to (a) provide a single datum for each object and (b) further minimize the effect of outliers in the area/ mass ensemble for a given object. This technique has proven to be a reliable means of estimating object area/mass ratios and masses when applied to fragmentations of space-craft and rocket bodies, space objects such as calibration spheres, and well-characterized objects, usually of simple geometric shape, used in internal calibration analyses conducted at NASA Johnson Space Center.

The EVOLVE 3.0 model (Reynolds, 1990, 1991) uses a model for the area-to-mass ratio that is anchored at the large end by the calculated area-to-mass ratios for intact satellites, and at the small end by assuming that the smallest fragments are thin plates with a density of the most common spacecraft material, aluminum.

The area-to-mass results are shown categorized by breakup in Fig. 6.2. A solid line that represents the EVOLVE model prediction is shown for each case. The Delta second stage breakups (*Landsats 1-3* and *Nimbus 6*) show a large amount of spread in area-to-mass ratios (Fig. 6.2a). There are a few dense pieces, but the majority are low-density objects with high area-to-mass values, as would be expected of insulation material. There also appears to be a population that falls near the theoretical curve, which may be indicative of plate-like pieces. The data (Fig. 6.2b) from the other rocket body explosions (*Transit 4A*/Able Star(t), Titan IIIC-4/Transtage, *OPS 7613* and *Nimbus 4*/Agena D, *SPOT-I*/Ariane H8, and *Fengyun 1-2*/Long March CZ-4) shows an even higher concentration of low-density objects with high area-to-mass values. In general, the higher area/mass portion of the distribution is dominated by the Ariane and Long March rocket body clouds. Aluminum construction figures prominently in the construction of both vehicles. Ariane debris in particular exhibit higher area/mass, perhaps indicative of the foam insulation utilized on the Ariane H8 third stage. Figure 6.2c, depicting the data from the Soviet antisatellite (ASAT) weapon tests (*Cosmos 252, 375, 397,* and *886*), resemble the data from rocket body explosions with somewhat lower average area-to-mass values. Note-worthy, however, is the clustering of massive objects, evidently the hulks of the ASAT interceptor vehicles. The P-78 (*SOLWIND*) test follows the theoretical curve most closely (Fig. 6.2d). The *Cosmos 1275* breakup (Fig. 6.2e) is something of a puzzle. A possible cause of the breakup was an on-board battery explosion. However, the size distribution of the fragments does not fit the pattern expected for that type of breakup. Instead, the size distribution closely matches that expected for a collision. For that reason, we have included this breakup in the collision class, noting that it is uncertain as to the exact cause of the breakup. Assuming that the *Cosmos 1275* breakup pieces are plate-like, their area-to-mass values are consistent with aluminum plates approximately 1 mm thick. The other breakups shown (*Cosmos 1813* and *1823*) show trends similar to those described above.

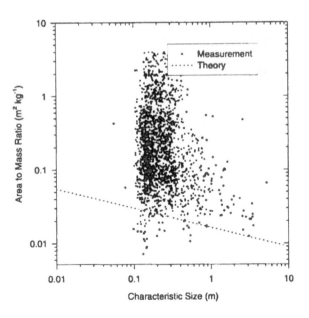

Fig. 6.2a. Area-to-mass ratios for delta breakups.

Other Rocket Body Explosions

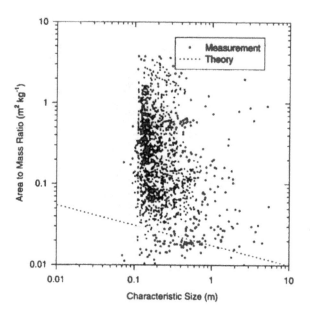

Fig. 6.2b. Area-to-mass ratios for other breakups.

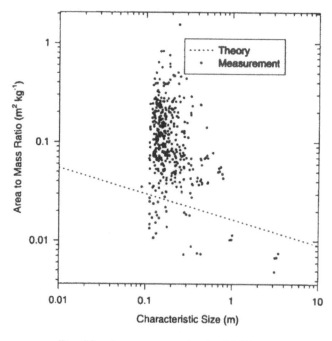

Fig. 6.2c. Area-to-mass ratios for ASAT test.

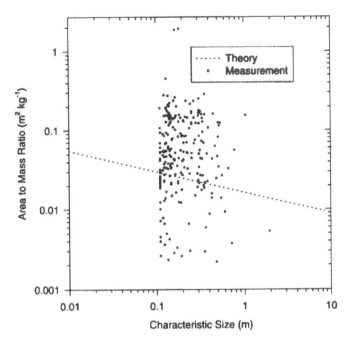

Fig. 6.2d. Area-to-mass ratios for Solwind P-78 fragments.

Kosmos 1275 Breakup

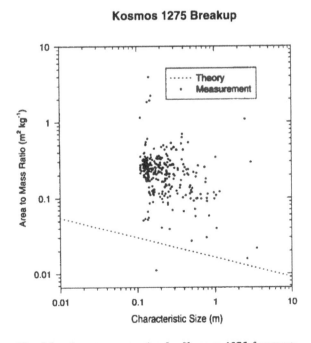

Fig. 6.2e. Area-to-mass ratios for Kosmos 1275 fragments.

6.4. Area-to-Mass Ratio for Debris from Laboratory Tests

Another useful comparison may be made with area-to-mass data for fragments produced in laboratory tests. Fragments from laboratory collision and explosion tests were weighed and their dimensions measured. For collisions, a sample set of pieces from the 6470 and 6472 hypervelocity impact tests (Bohannon, Dalquist and Pardee, 1991) and debris produced during the Satellite Orbital Debris Characterization Impact Test (SOCIT) (Hogg, Cunningham and Isbell, 1993) were used. For explosions, 8 pieces from the European Space Agency (ESA) Ariane tests (Fucke, Sdunnus, 1993) were used. In addition to the mass of the pieces, three axis lengths were measured: the longest axis (a), the longest axis perpendicular to the first (b), and the axis perpendicular to the other two (c). These three axes are averaged to obtain a characteristic sphere diameter: $d = [a + b + c]/3$.

Support for this procedure was found in the radar cross-section work reported by Sato *et al.* (1994). They found that the radar cross section profiles of debris objects measured by the MU upper atmosphere radar in Japan could be explained by approximating the shapes of these debris objects as ellipsoids. For each debris piece measured, the three-axis size measurements described above were used to calculate the average projected area of an equivalent ellipsoid having three axes of the same length as the debris piece. The characteristic sphere method gives areas within 20% of the values computed by the equivalent ellipsoid method. The similarities of these two methods increase our confidence that the approximation of the average area based on a characteristic sphere does not introduce large errors.

We computed area-to-mass ratios ξ for the laboratory fragments by dividing the spherical cross-sectional area by the measured masses. Figure 6.3 compares the results for the orbiting debris (all breakups combined) with the laboratory measurements of breakup fragments, and with the standard theoretical (EVOLVE 3.0) curve. Note that the collisional debris pieces have a much larger spread in the area-to-mass ratios than the scaled Ariane test pieces. The Ariane pieces are very plate-like and exhibit area-to-mass ratios relatively independent of size. Because both the average area and mass of a plate is proportional to the square of its size, the area-to-mass ratio is approximately constant, depending only on the thickness and density of the material in the plate (at least for plates with lengths significantly larger than their thickness). If measurements of the orbiting debris of a particular breakup show a similar area-to-mass independence from size, then it is possible to assume the debris is plate-like and to approximate the thickness and/or material composition of the debris.

6.5. Mass Distributions for Orbital Debris Objects

The masses for debris objects were determined by dividing the area by the area-to-mass ratio as determined above. Areas were determined from radar cross-sections (RCSs) by the procedure described earlier.

Cumulative mass distributions were calculated for each breakup by dividing the range of masses into 20 equal logarithmic ranges, and then calculating the number of fragments

Area-to-Mass Ratios

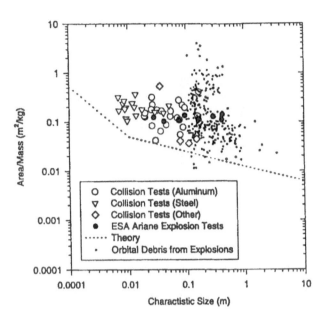

Fig. 6.3. Area-to-mass ratios for laboratory explosions.

with masses equal to or larger than each of the 20 steps. We have chosen to compare the observed mass distributions with those predicted by the models used in the EVOLVE 3.0 (and many other contemporary) orbital debris environment code (Reynolds, 1990, 1991). The breakup model used by EVOLVE is based on the classic work by T.D. Bess (Bess, Dale, 1975), who analyzed the cumulative fragment distributions for hypervelocity impacts, high-, and low-intensity explosions.

The breakup models in the EVOLVE 3.0 long-term debris environment evolution computer model (Reynolds, 1990, 1991) address three cases: hypervelocity impact, low intensity explosion, and high intensity explosion. Low intensity explosions are considered those caused by, for example, vessel overpressurization causing a rupture, followed by the subsequent destruction of the remaining hulk by acoustic 'ringing' of the structure. High intensity explosions are considered those in which an explosive was in direct contact with the spacecraft structure. In all cases, the EVOLVE 3.0 model conserves mass by normalizing the cumulative mass of debris objects greater than 1 mm in diameter to the estimated dry mass of the fragmenting body.

The cumulative distribution of masses from a hypervelocity impact is assumed to be governed by a power law:

$$N(m) = 0.4478(m/m_e)^{-0.7496}, \qquad (6.1)$$

where N is the number of objects of mass m (grams) and greater and m_e (grams) is the mass of ejecta created by the impact.

The cumulative distribution of masses from explosions are governed by an exponential proportional to $\sqrt{m}$. For low-intensity explosions, the distribution is described by:

$$N(m) = 1.707 \times 10^{-4} \; m_t \; \exp(-0.0206 \sqrt{m}), \; m \geq 1936 \; g \qquad (6.2)$$
$$= 8.692 \times 10^{-4} \; m_t \; \exp(-0.0576 \sqrt{m}), \; m < 1936 \; g \qquad (6.3)$$

In the case of a high intensity explosion, the target mass m_t is modified such that 90% of the mass goes into the exponential form, whereas the remaining 10% goes into a power law (slope = –1) between 1 mm and 1 cm. This *ad-hoc* measure was taken to account for the small particles expected to be produced in the vicinity of the explosion. Otherwise, the exponential forms described by Equations 2 and 3 would not produce the number of small particles required by *in-situ* and ground-based radar measurements of the space environment.

For breakups judged to be low- or high-intensity explosions, the cumulative number of orbital debris fragments were plotted on a logarithmic axis against the square root of fragment mass. For hypervelocity impact breakups, the logarithm of the cumulative number was plotted against the logarithm of fragment mass. For each case, data points derived from the 1994 version of the EVOLVE 3.0 orbital debris environment model are also plotted. The EVOLVE 3.0 data were obtained for the same epoch as the last data set for which fragment masses were calculated from orbit decay. This is an important point,

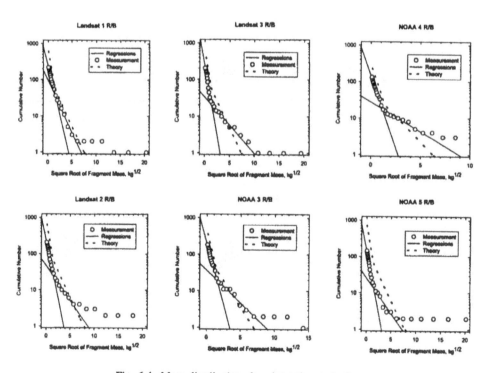

Fig. 6.4. Mass distributions low intensity explosions.

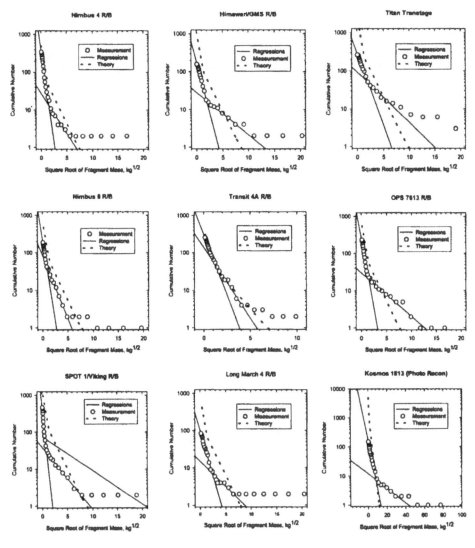

Fig. 6.4. Continued.

since it is necessary to take into account the loss of debris objects by decay and reentry, and the EVOLVE 3.0 model takes this into account. Results are shown for low-intensity explosions (mostly upper stage breakups) in Fig. 6.4, for high-intensity explosions (anti-satellite tests) in Fig. 6.5, and for a known hypervelocity impact breakup (SOLWIND) in Fig. 6.6.

Study of the plots for low-intensity explosions shows that the observed distributions differ from the model distributions in important ways. In most cases, the model predicts many more fragments than were actually observed. This is apparently because many of the progenitor objects break up into several large fragments, only one of which fragments

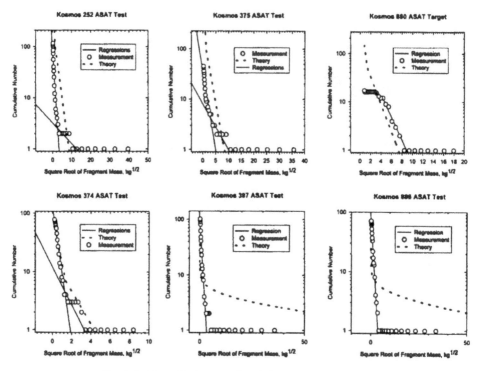

Fig. 6.5. Mass distributions for high intensity explosions.

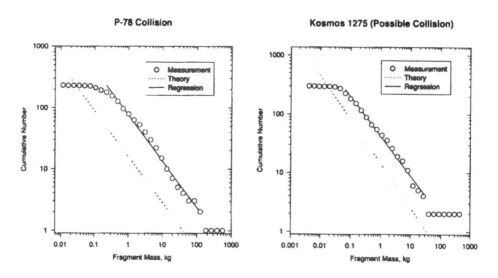

Fig. 6.6. Mass distributions for collision breakups.

Table 6.1. Least-squares fits to mass distributions

	Large Debris Intercept	Slope	Small Debris Intercept	Slope	Break Point Mass, kilograms
Low Intensity					
Explosions					
Landsat 1	1.9783	−0.2864	2.3782	−0.5467	2.3311
Landsat 2	1.6730	−0.1924	2.3581	−0.6442	2.2593
Landsat 3	1.5209	−0.1453	2.3887	−0.7780	2.3034
NOAA-3	1.5620	−0.1786	2.3027	−0.6817	2.2231
NOAA-4	1.4379	−0.1619	2.1619	−0.7849	1.0804
NOAA-5	1.4100	−0.1951	2.1791	−0.7139	2.3165
Nimbus-4	1.4450	−0.2131	2.6234	−0.9915	2.4822
Nimbus-6	1.9210	−0.3343	2.3424	−0.8645	3.1354
Himawari	1.4556	−0.1126	2.2779	−0.5414	3.2946
Transit 4 A	2.1456	−0.3779	2.4983	−0.7551	2.0036
Titan Transtage	1.9717	−0.1331	2.4743	−0.3822	2.5973
OPS 7613	1.4859	−0.1177	2.4133	−0.7681	2.8210
Spot Viking	1.5907	−0.1678	2.6646	−0.1678	1.8512
Fengyun 1–2	1.1793	−0.1348	2.0087	−0.4828	4.2095
Kosmos 1813	1.0845	−0.0253	2.2009	−0.1513	80.5542
High Intensity					
Explosions					
Kosmos 252	0.6636	0.0725	1.9511	−0.5792	5.3310
Kosmos 374	0.8803	−0.1849	2.1149	−1.1364	1.1866
Kosmos 375	0.6711	−0.0538	1.5522	−0.3124	14.7425
Kosmos 397			2.0026	−0.6197	9.5419
Kosmos 880			1.2575	−0.0347	19.2396
Kosmos 886			2.1120	−0.2379	12.3996
Collisions					
Solwind P78			1.8783	−0.7516	126.7500
Kosmos 1275			1.6218	−0.6779	26.9100

completely in a way similar to that expected from the model. For the Cosmos antisatellite tests, agreement between observation and model is relatively good. However, for these tests, the EVOLVE 3.0 model assumed that only 10% of the satellite mass participated in the breakup. For the only case acknowledged to be the result of hypervelocity impact (SOLWIND), the EVOLVE 3.0 model predicted substantially less debris than actually found. A reasonable explanation of this difference is that the orbital decay part of the EVOLVE model has predicted the fragments to reenter more rapidly than actually has occurred. It may be useful for further analysis to have available the least-squares fits to the mass distributions for these breakups. Table 6.1 provides the least-squares fits to all the breakups, and Table 6.2 provides average values and standard deviations for the three classes of breakups.

Table 6.2. *Average least-squares fits to mass distributions*

	Low-Intensity Explosions	High-Intensity Explosions	Collisions
Large Debris Intercept	1.6339	0.8249	
Standard Deviation	0.2663	0.2001	
Large Debris Slope	−0.1936	−0.0479	
Standard Deviation	0.0793	0.1062	
Small Debris Intercept	2.3580	1.8845	1.7501
Standard Deviation	0.1707	0.3481	0.1814
Small Debris Slope	−0.6402	−0.4388	−0.7148
Standard Deviation	0.2075	0.3744	0.0521
Mass at Slope Change	2.5884	2.3089	
Standard Deviation	0.5927	1.0893	

It is informative to compare the total mass of fragments involved in a breakup with the EVOLVE model prediction. For this, we need to determine how much of the progenitor object is involved in the breakup. The most straightforward case is that of the Cosmos ASAT series, where it is assumed that the main satellite fires an explosive charge, and releases a cloud of fragments. The main satellite remains more or less intact. A reasonable guess is that only about 10% of the mass ends up as fragments, and this was assumed for the EVOLVE model predictions, as noted above.

It is less straightforward for some of the other breakups, but we can use the EVOLVE breakup model as a guide to estimating what part of the object participated in the breakup to small particles. The EVOLVE breakup model for explosions predicts that the mass of the largest fragment will be less than about 10% of the initial mass of the intact progenitor object. We found several breakups where the observed mass distribution did not follow this pattern. For these, the largest masses were of the order of 1/2 to 1/3 of the progenitor mass. Evidently, satellite breakups generally follow a more complex sequence than implied by the simple model. One or more large fragments of near-equal size are generated, plus a number of small fragments that follow approximately the breakup model. For these cases, in order to estimate the mass involved in the part of the breakup that generates the small particles, we discarded the largest one or two masses, if they were larger than about 4 times the next smallest fragment.

The total fragment mass and the estimated mass involved in fragmentation into small particles are listed in Table 6.3 for low-intensity, high intensity, and hypervelocity collision breakups. For the low-intensity rocket body explosions, the actual mass of debris generated by the breakup is significantly less than the mass available from the progenitor satellite, and are also less than that predicted by the model. For the high-intensity antisatellite explosions, there is reasonably good agreement between the observed total and the modeled total. As noted above, the actual total mass of small debris from the hypervelocity impact cases is significantly larger than predicted by the model.

Table 6.3. Fraction of total satellite mass in breakup fragments

Satellite launch	Breakup	Original Mass, kg	Total Fragment Mass, kg	Total Breakup Mass, kg	Breakup Mass Fraction
Low Intensity Explosions					
1972–058 Landsat 1	Delta R/B	794	797	382	0.48
1975–004 Landsat 2	Delta R/B	872	946	621	0.71
1978–026 Landsat 3	Delta R/B	872	729	352	0.40
1973–086 NOAA 3	Delta R/B	794	539	227	0.29
1974–089 NOAA 4	Delta R/B	872	338	338	0.39
1976–077 NOAA 5	Delta R/B	872	913	531	0.61
1970–025 Nimbus 4	Agena R/B	700	667	385	0.55
1975–052 Nimbus 6	Delta R/B	872	822	251	0.29
1977–065 Himavari	Delta R/B	872	1370	401	0.46
1961–015 Transit 4 A	Ablestar R/B	789	488	488	0.62
1965–082 OV 2 1/LCS 2	Titan Transtage	3130	3080	2310	0.74
1969–082 OPS 7613	Agena R/B	1588	793	514	0.32
1986–019 Spot Viking	Ariane R/B	1400	1612	257	0.18
1990–081 Fengyun 1–2	Long March 4 R/B	1000	958	137	0.14
1987–004 Kosmos 1813	Reconn Satellite	6000	10350	2151	0.35
				Average	0.44
High Intensity Explosions					
1968–097 Kosmos 252	ASAT Test	1400	1680	120	0.09
1970–089 Kosmos 374	ASAT Test	1400	110	110	0.08
1970–091 Kosmos 375	ASAT Test	1400	1414	144	0.10
1971–015 Kosmos 397	ASAT Test	1400	1340	70	0.05
1976–120 Kosmos 880	ASAT Target	650	695	373	0.57
1976–126 Kosmos 886	ASAT Test	1400	1219	79	0.06
				Average	0.16
Collisions					
1979–017 Solwind	P-78 ASAT Test	850	1163	410	0.48
1981–053 Kosmos 1275	Comm Satellite	800	1015	99	0.12
				Average	0.30

6.6. Conclusions and Implications

We find that the area-to-mass ratios for objects in a given swarm of debris from an on-orbit breakup display considerable scatter, not unexpected for fragments from an explosive event. However, in nearly all cases, the mean area-to-mass ratio of the objects in the debris cloud objects is significantly greater than that predicted by the 1994 EVOLVE 3.0 model. As a consequence, the debris fragments will decay from orbit more rapidly than predicted by the model. This will result in an over-prediction of future debris populations by this model, at least for the sizes encompassed by the measured data.

Area-to-mass ratios can be used to determine mass densities of the debris objects by assuming a geometry for the objects (plate, sphere, *etc.*) as described in reference (Anz-Meador, Rast and Potter, 1993). Any choice of geometry will lead to a low mean mass density (relative to the 1994 EVOLVE 3.0 assumptions) and a wide distribution of mass densities. This result poses a problem for determining the correct amount of shielding to be used to protect spacecraft against hypervelocity impact by orbital debris. The current practice is to assume that the debris objects are spheres with mass densities approaching that of aluminum as size is decreased. It is clear that this is an oversimplification, although it may represent the real situation well enough. Further study and analysis is needed to determine if this is the best assumption.

We also find that in most cases, the satellite or rocket body does not break up in the way assumed by some environment models, for example, EVOLVE as it was formulated in 1994. In most cases, only a portion of the total mass yields a fragment size distribution described by the breakup model, whereas the 1994 EVOLVE 3.0 model assumed that the entire mass of the object breaks up into a size distribution described by the breakup model. The implication of this result is that the population of small debris may be overestimated by such models.

6.7. Acknowledgement

The authors gratefully acknowledge the provision of historical USSPACECOM two line element sets ("elsets") by Mr. Nicholas L. Johnson, Orbital Debris Project Scientist, of NASA Johnson Space Center. The analyses discussed in this work would have been impossible without access to this large data set.

CHAPTER 7

SPACE DEBRIS PRODUCTION IN DIFFERENT TYPES OF ORBITAL BREAKUPS[*]

N.N. Smirnov, V.F. Nikitin and A.B. Kiselev

The paper investigates orbital breakups as a major source of space debris production. Different types of breakups under the influence of nonuniform internal loadings are regarded caused by internal explosions or hypervelocity collisions with debris particles. The worked out new thermodynamic criterion of destruction based on the critical value of energy dissipation in irreversible transformations makes it possible to determine the number and fluxes of fragments formed in breakups. The comparative analysis shows that distribution functions of the number of fragments versus mass have extrema strongly depending on peculiarities of breakup scenario that can produce essential difference from forecasts given by the accepted fragmentation models. Breakups caused by chemical explosions or hypervelocity particles impacting pressurized vessels produce different fluxes of fragments.

7.1. Introduction

Since the time the scientific community faced the orbital debris problem there arose a question of its impact on the Space environment (Chobotov, 1990; Flury, 1992; Loftus, Anz-Meador and Reynolds, 1992). Collisions and orbital breakups were considered to play the most important role in determining the scenario of debris evolution in low Earth orbits (Kessler, 1991) and in geostationary orbit (Yasaka, Ishii, 1991).

---
[*] The present research was supported by the Russian Foundation for Basic Research (grants 00-15-99060; 00-01-00245 and 98-01-00218).

The description of debris evolution and the determination of the amount of space debris in elliptical orbits needs adequate modeling of the major sources of debris production and consumption: in-orbit explosions, hypervelocity collisions, operational debris separation, slowing down of fragments captured by the atmosphere (Nazarenko, 1993; Smirnov *et al.*, 1993; Chobotov, Spencer, 1991; McKnight, Nagl, 1993).

Operational debris originates mostly in separations of satellites and last stages of rocket-carriers and the mean rate of its production is assumed to be 4 objects per launch (Talent, 1990), that gives about 500 objects per year. The same amount of fragments can appear in one orbital breakup being the result of an explosion.

Thus fragmentations of satellites due to different reasons and mostly explositions of the upper stages are supposed to give the most essential contribution to the space debris production (Loftus, 1987). It was only in 1981 when Don Kessler, NASA-JSC was able to correlate space debris from satellite breakups recorded by NORAD/ADCOM to upper stages of rocket carriers left in orbit after completion of their mission (NASA-JSC Technical Memorandum, 1981). Since 1969 up to 1981, ten cases of breakup of Delta second stages left in orbit after mission took place (Webster, Kawamura, 1992). The duration of stay of the vehicles in orbit before the explosion varied from 1 day up to 5 years.

One of the most probable potential causes of orbital breakups is fuel and oxidizer tanks overpressurization and fracture of the common bulkhead allowing mixing and ignition of the residual propellants that would most probably result in an explosion (McDonnel Douglas Astronautics Company, 1982).

Theoretical and experimental investigations of breakups of fuel tanks showed that the results of breakups: number, masses and velocities of fragments — strongly depend on the peculiarities of the process of energy release inside the tank (Smirnov, Nikitin, Kiselev, 1996; Smirnov, Lebedev, Kiselev, 1994; Fucke, 1993).

The other probable potential causes of orbital breakups are hypervelocity impacts on pressurized compartments and chemical explosions. Gas-filled and fluid-filled vessels show different behavior in breakups caused by collisions and internal loadings.

Theoretical and experimental modeling of hypervelocity collisions of debris fragments with pressurized vessels is important not only from the point of view of debris production, but also for studying of orbital fragments collisions impact on space vehicles and conceptual designing of new shields (Christiansen, Horn and Crews, 1990; Christiansen, 1992).

The aim of the present investigation is to create physical and mathematical models of breakups able to describe peculiarities of the fragmentation phenomena for different breakup scenario and to determine the number, mass and velocity of fragments.

The present chapter contains the description of the physical model of breakup events (section 7.2). The detailed mathematical models of energy release in internal chemical explosions, dynamical loading, deforming, accumulation of damages and breakup of thin-walled structures are described in the Annex. For a more detailed explanation one could get acquainted with some other publications by the authors (Smirnov *et al.*, 1996; Smirnov *et al.*, 1994; Smirnov, Panfilov, 1995; Smirnov, Tyurnikov, 1994; Smirnov *et al.*, 1997; Kiselev, Yumashev, 1990; Kiselev, 1994; Kiselev, 1997). The results of fragmentation for different scenario of energy release and comparisons with the experiments are discussed in sections 7.3–7.5.

7.2. Physical Models of Breakup Processes

On regarding different possible breakup scenario one can come to the following classification of breakup types.

1. Breakups due to chemical explosions. Breakups of this type result from internal loading due to chemical energy release inside the structures.
2. Breakups due to overpressurization — occur in gradual uniform loading of the closed containments resulting from physical changes of the internal media: heating, evaporation, etc.
3. Breakups due to collisions — occur in incidental loading of the structure resulting from the transformation of the kinetic energy of the impactor.

Each type of breakups also has a variety of possible scenario. Chemical energy release in premixed combustible materials contacting the spacecraft structure can take place in two modes: deflagration (or slow combustion) and detonation (or quick-going process when the rate of reaction zone propagation has the order of magnitude of kilometers per second) (Smirnov, Panfilov, 1995; Smirnov, Tyurnikov, 1994). The rates of wall loadings and maximal loads differ several orders of magnitude for deflagration and detonation resulting in different fragmentation patterns, masses, velocities of fragments (Smirnov et al., 1997).

The nonpremixed combustible systems (hypergolic propellants) reacting in an oscillating diffusion mode (Smirnov et al., 1996; Smirnov, Nikitin, 1997), produce nonuniform wall loading resulting in formation of a very wide spectrum of fragments distribution versus mass with several extrema (Smirnov et al., 1997).

Collisions of pressurized vessels with hypervelocity particles can result in different breakup scenario for fluid-filled and gas-filled containments.

Generally the breakups are classified due to their "intensity" i. e. the number of fragments versus mass distribution function as "high-intensity explosions", "low-intensity explosions" and "collisions". This classification mostly deals with available empirical formulae for fragments production modeling (Chobotov, Spencer, 1991; Su, Kessler, 1985) and does not reflect the physics of the fragmentation phenomena. Different fragmentation scenario we discussed can produce a greater variety of fragments distribution functions. Besides as it was discovered in (Potter, Anz-Meador et al., 1994; Anz-Meador, Potter, 1996) the existing models for fragments distribution cannot describe adequately the peculiarities of the real breakup events. Below we shall give a brief description of models for several breakup scenario.

The overall variety of in-orbit breakups contain the following characteristic stages:

- energy release and loading of the structural elements of the satellite;
- dynamical deforming of the structure and accumulation of damages;
- destruction of walls, cracks' growth, fragments formation and dispersion.

7.2.1. The Model for the Energy Release

The internal loading of the structure can take place due to chemical energy release (deflagration and/or detonation) or due to kinetic energy transformation in hypervelocity collisions.

In case of chemical explosion the reactants (in propellants tanks, for example) can be in different phases: liquid, gaseous, solid. Thus the mathematical model for multiphase chemically reacting media should be applied.

In case of hypervelocity collision the fragmentation of the impactor and the satellite's structures in the collision zone gives birth to formation of a cloud of small hypervelocity fragments entering the pressurized gas- or fluid-filled containments. The rapid slowing down (and possible combustion in the oxygen containing atmosphere) of those fragments brings to the localized energy release inside the containment similar to an explosion. Then the dynamics of the internal wall loading depends on the density of the released energy and compression waves propagation and reflections inside the containment.

Those processes of energy release and "point blasts" can be also described with the help of the mathematical model for multiphase chemically reacting flows (Smirnov *et al.*, 1996; Smirnov, Zverev, 1992). A brief description of the model and numerical scheme is given in the Annex.

a) Combustion of polydispersed sprays in weightlessness

On ignition of the residual fuel in the tanks of a spacecraft after an accidental breakup of the common bulkhead separating fuel and oxidant the polydispersed mixture in the tank is in an unknown pattern of dispersal. Thus the energy release in combustion does not depend on the total mass of the residual fuel, but rather on the mass of components entering into reaction. The following examples illustrate peculiarities of combustion of polydispersed sprays containing fuel in a condensed phase and oxidant in the gaseous phase. Other compositions (gaseous fuel-liquid oxidant) could be also treated with the help of the multiphase model, discussed in the Annex.

The investigation of sprays combustion was performed for model multicomponent fuel particles permitting both gas-phase and heterogeneous reaction modes.

The numerical simulations of ignition and combustion were performed for polydispersed mixtures wherein the initial particles' radius had a stochastic scatter with the following probability density function

$$f(r) = \begin{cases} \dfrac{2(r-r_{min})}{(r_{med}-r_{min})(r_{max}-r_{min})}, & r_{min} \le r \le r_{med}; \\[2mm] \dfrac{2(r_{max}-r)}{(r_{max}-r_{med})(r_{max}-r_{min})}, & r_{med} \le r \le r_{max}; \\[2mm] 0, r < r_{min} \text{ or } r > r_{max}. \end{cases}$$

The particles regarded here contain not only evaporating components burning in the gaseous phase, but also solid component that could react in the surface only (heterogeneous reaction).

Figures 7.1 (a–c) illustrate the dynamics of the reaction zone in a cyclindrical vessel with a vertical axis of symmetry. The reaction zone is tracked here as the zone of maximal rates of volatiles oxidizing.

Figures 7.2 (a–c) illustrate the particles temperature for the corresponding times and show even the greater width of the reaction zone. Some particles in the zone of reaction products do not burn out to the end due to the lack of an oxydant.

Temperatures of particles here correspond to those of the solid component liquid volatile components evaporate earlier at lower temperatures.

The results shown in Figs. 7.1–7.2 were obtained for the polydispersed mixtures of particles with the air. The particles had the following characteristic diameters:

$$d_{min} = 10^{-5} \text{ m}; \ d_{med} = 5 \cdot 10^{-5} \text{ m}; \ d_{max} = 7 \cdot 10^{-5} \text{ m}.$$

The other set of numerical simulations was performed for polydispersed particles mixtures with pure oxygen. The spray was characterized by the presence of larger particles:

$$d_{min} = 10^{-5} \text{ m}; \ d_{med} = 5 \cdot 10^{-5} \text{ m}; \ d_{max} = 10^{-4} \text{ m}.$$

The total list of problem parameters is very large and thus not given in the present paper. The dynamics of the reaction zone (gas phase reaction intensity) and the particles temperatures are shown in Figs. 7.3–7.4. It is clearly seen that the width of the reaction zone is much larger for the present case and irregularities present in the reaction zone are much stronger. That is mostly due to the fact that the larger particles have a longer induction time for ignition and longer combustion time. Large particles due to the large content of volatiles cause nonuniformities in the volatiles concentration field promoting the formation of hot spots in the reaction zone (Fig. 7.3).

The increase of the initial content of oxygen brings to widening the reaction zone due to the continuation of heterogeneous combustion of particles after the gas-phase reaction being terminated.

The results show the strong influence of the burning rates and peculiarities of energy release on the overall modeling of spray combustion and the rates of pressure growth. Those peculiarities produce qualitative effects on the wall loading pattern and the breakup scenario.

b) Diffusive combustion of hypergolic propellants in accidental mixing of components on perforation of the common bulkhead

The incidental mixing and ignition of the combustible components initially preserved in different containments separated by a bulkhead can provide the possibility for an explosion. Hypergolic propellants are among the most dangerous ones from this point of view since they have a very low ignition temperature. Thus no ignition source is necessary and chemical reaction starts at first contact of the components. The large rate of chemical reaction as being compared with the rate of diffusive mixing leads to establishing

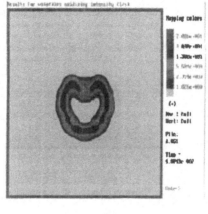

a

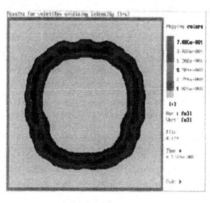

b

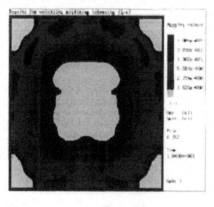

c

Fig. 7.1. The dynamics of the reaction zone in polydispersed fuel-air mixtures combustion.

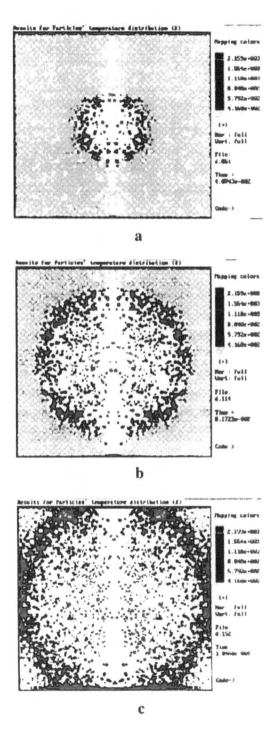

Fig. 7.2. The particulate phase distribution and temperatures for successive times in polydispersed fuel-air mixtures combustion.

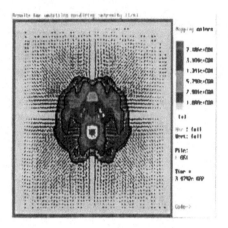

a

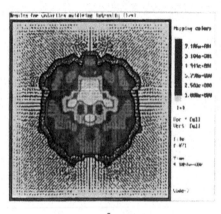

b

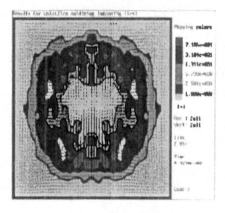

c

Fig. 7.3. The dynamics of the reaction zone in polydispersed fuel-oxygen mixtures combustion.

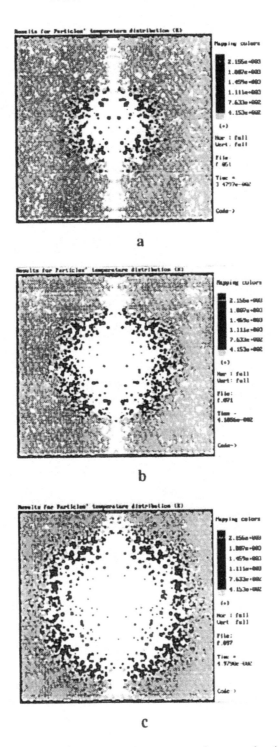

Fig. 7.4. The particulate phase distribution and temperatures for successive times in polydispersed fuel-oxygen mixtures combustion.

conditions under which the reactants are unable to penetrate deep through a flame not entering into reaction. The main limiting factor determining the rate and amount of energy release in the explosion thus is not the kinetics of reactions but rather the rate of diffusive mixing of the components.

The rate of diffusive mixing is rather low and diffusive damping of transport of reactants by the originating reaction products is usually present. Thus the rate and amount of energy release in diffusive combustion could not be considered dangerous. Nevertheless turbulence produced in the flowfield and in the flame zone increases the mixing rate by several orders of magnitude and can provide the possibility of secondary ignitions thus increasing the amount of released energy to dangerous quantities.

The rate of diffusive combustion determines the rate of pressure increase inside the vessel. The pressure increase up to a certain value leads to a breakup and expansion of reaction products and unburned fuel and oxidizer.

Thus, the main peculiarity of the breakup caused by the explosion of mixing hypergolic propellants is that the total amount of energy released by the time of breakup depends on dynamics of mixing and the rate of diffusive combustion and can be practically independent of the total amount of fuel and oxidant.

The aim of the present investigation is to determine the dynamics of turbulent diffusive flame propagation in homogeneous combustible gaseous systems wherein fuel and oxidant are not premixed. The two-chamber vessels are regarded separated by a bulkhead. At the initial instant fuel and oxidant occupy different chambers. Thus combustion starts under the conditions of initial gradients of concentrations and the developing turbulence plays an important role not only in the combustion process but as well in the process of mixture formation.

The rate of energy release responsible for loading and breakup of walls of containments in gas explosion is determined by turbulent mixing and diffusive combustion of components after destruction of a bulkhead.

After a complete or partial destruction of the bulkhead that can be modelled by opening orifices of different diameters the existing difference of pressures in the chambers can cause formation of a jet of one reactant into the chamber filled with the other reactant. The flowfield, the intensity of turbulence and the combustion rate depend on diameter of the orifice in the bulkhead and initial pressures in the chambers.

The mathematical model of the process is based on the modified $K - \varepsilon$ model accounting for the near-wall damping effect by making use of the Lam-Bremhorst low Reynolds model. The location of diffusion flame in its unsteady evolution is tracked by the isolines of reaction intensity.

We regard the unsteady motion and mixing of gaseous components in a cylindrical vessel having the form shown in Fig. 7.5. At the initial instant the reactants are separated by a common bulkhead. The main parameters: pressures, temperatures, densities — are uniform in both containments, but pressures can be different. A hole of a definite area is opened in the bulkhead in the axis of symmetry instantaneously at time $t = 0$, that causes formation of compression and rarefaction waves in both containments. A turbulent jet of components penetrates the low pressure chamber and rarefaction wave propagates in the high pressure chamber.

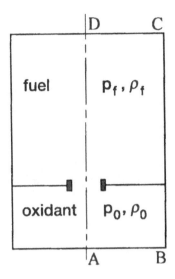

Fig. 7.5. A schematic picture of a propellant tank with a common bulkhead separating fuel and oxidant compartments. An axial orifice simulates a perforation in the bulkhead allowing mixing the components.

Mixing of components initiates combustion in a low pressure chamber that leads to a rapid pressure increase. Thus, main governing parameters of the process are the initial pressure ratio in both tanks and the size of the hole.

To trace the dynamics of the reaction zone it is necessary to visualize the reaction intensity $\dot{Y}$. Maximal values of $\dot{Y}$ correspond to maximal reaction rates in the flow, thus, showing the location of the flame zone.

Figures 7.6a–f show the reaction intensity isolines for different times. Initially reaction zone is located just near the orifice (Fig. 7.6a). The scale for reaction intensity values (the upper right corner of the figures) is logarithmic. That is why the light grey lines relate to a very small reaction intensity (of the order 10^{-8} that is six orders less than maximal reaction intensities within the flame zone). Thus, the flame zone practically is rather narrow and it is located just in the orifice (Fig. 7.6a) bounded by the dark black isolines of a higher reaction intensity. Turbulent diffusion increases the width of the flame zone. The flame zone is removed from the orifice by the gas flow from the high pressure tank. The flame is bent and the unsteady state jet diffusion flame occurs (Figs. 7.6b–c). The maximal reaction intensity happens to be at first at the nozzle of the jet (Fig. 7.6a), due to the more close contact of oxidant with fuel. Later the reaction intensity increases also around the orifice (Fig. 7.6b) due to the fact that fuel is sucked there by a flow induced by the turbulent jet. Later ($t = 3.5 \cdot 10^{-3}$ s $\div 4.5 \cdot 10^{-3}$ s), maximal reaction intensity is sustained only around the orifice and the flame zone in the nozzle of the jet is separated from the diffusion flame near the opening (Fig. 7.6c). The reason for such a separation is in the rapid increase of the amount of reaction products diluting the mixture of fuel and oxidant and preventing them from coming together to form a combustible mixture. The

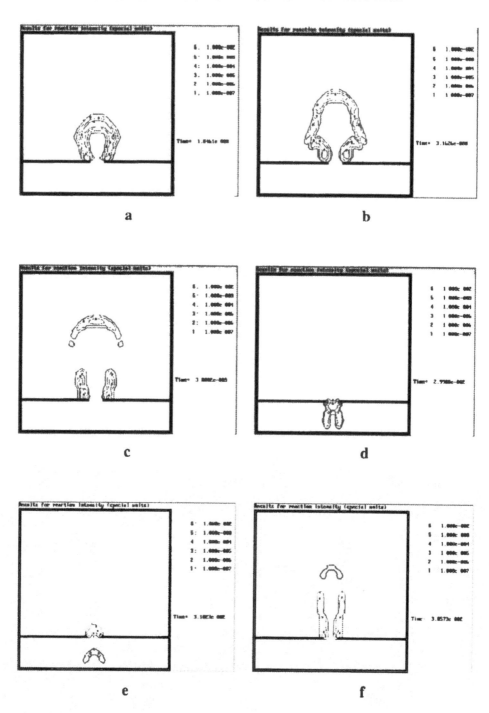

Fig. 7.6. The reaction zone evolution in unsteady-state diffusion combustion taking place in a two-chamber vessel after opening an orifice in the common bulkhead. The isolines of reaction intensities are shown for successive times a, b, c, d, e, f.

gradients of concentrations of reagents decrease in this zone and diffusion fluxes turn to be too low to sustain combustion in spite of a high level of turbulence. The diffusion fluxes (and concentration gradients) are rather high near the orifice where the jet starts. These results in a small vicinity of the orifice coincide with the existing solutions for diffusion combustion in a boundary mixing layer. Those solutions include a singularity in the angle point with infinite radial fluxes.

The numerical solution for viscous turbulent flow shows that in time the influence of the wall (the common bulkhead) lowers down the fluxes and flame extinguishes. Slowing down the turbulent jet for the times $t \approx 5 + 8 \cdot 10^{-3}$ s also causes the decrease of turbulent diffusion fluxes.

By the time $t \approx 2.5 \cdot 10^{-2}$ s pressure in the upper (fuel) tank surpasses that in the lower (oxidant) tank due to energy release in combustion and expansion of the reaction products. Under the influence of the pressure gradient there starts a counterflow of gas mixture from the upper to the lower tank. The reversive jet induces a torroidal vertex in the lower tank and oxidant is sucked to the beginning of the jet near the orifice. The increase of concentration gradients in this zone and the growth of diffusion fluxes gives birth to a reaction zone around the jet near the orifice (Fig. 7.6d).

The further development of the flow inside the oxidant tank leads to a separation of flame zone from the flame stabilized near the orifice (Fig. 7.6e). By that time pressure in the oxidant tank surpasses that in the upper tank due to combustion and the gas flow from the lower to the upper tank starts once again. The diffusion flame stabilized near the orifice is pushed out of the oxidant tank by the gas flow and forms a diffusion flame in the upper (fuel) tank (Fig. 7.6f).

The development of combustion process in the upper tank repeats the scenario of the previous development of the process in this tank with the only difference that now, due to the presence of the essential amount of reaction products inside the tank, we do not have intensive combustion at the nozzle of the jet but only intensive combustion in the vicinity of the orifice.

The pressure fields evolution within the double compartment structure makes it possible to follow the dynamics of nonuniform loading of the walls of the vessel.

c) Detonation of the mixture in propellant tanks

Combustion of fuel-oxidant mixtures could take place in two modes: deflagration (or normal combustion) and detonation. The velocities of normal deflagration wave propagation are rather low from centimeters per second for laminar flames to meters per second for turbulent. Detonation waves due to a different mechanism of their propagation have much higher velocities — kilometers per second. As it was shown both theoretically (Smirnov, Panfilov, 1995) and experimentally (Smirnov, Tyurnikov, 1995) for hydrocarbon fuel-oxidant mixtures deflagration to detonation transition could take place under certain conditions.

One of the typical scenario of the deflagration to detonation transition is illustrated in the Fig. 7.7. The figure shows the successive pressure profiles calculated for a plane detonation wave propagating in a cylindrical tube from the left to the right. The curves

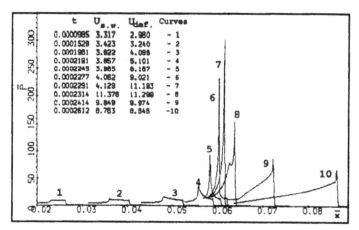

Fig. 7.7. Successive pressure profiles illustrating the dynamics of the deflagration to detonation transition process in combustible gaseous mixtures. The waves propagate from the left to the right hand side.

marked by higher number, correspond to increasing times. Curves 1 and 2 show the pressure profiles for the deflagration wave propagation. Expansion of the heated reaction products brings to a formation of a leading shock wave ahead of flame. The shock wave sets the initially quiescent gas into motion and causes the pressure increase. Then comes the flame zone wherein the energy release due to chemical transformations takes place accompanied by a slight decrease of pressure. Acceleration of the flame zone brings to a formation of an additional compression wave ahead of it (curves 3, 4, 5). The pressure and temperature increase ahead of the flame promotes chemical transformations and brings to a formation of a strong detonation wave (curves 6, 7) overtaking the leading shock wave (curve 8). The so-called overdriven detonation wave formed in the gas, slows down gradually to a self-sustaining detonation mode (curves 8, 9, 10). The chemical energy release takes place just after the pressure increase in the detonation wave.

It is seen from the Fig. 7.7, that the maximal pressures are essentially different for detonation, deflagration and the deflagration to detonation transition under the conditions of one and the same energy release per fuel mixture mass unit. But the duration of the high pressure pikes is lower the higher is the pike. Those waves being reflected from the wall of a containment could produce different loading patterns depending on the combustion mode. Thus the maximal loading pressure and the time of loading could vary within the range of two orders of magnitude. The different loading patterns could produce different effects on the walls thus bringing to different breakup scenario.

7.2.2. The Model for the Dynamical Deforming of Walls

The mathematical models for the dynamical deforming of walls of containments under the influence of internal loading were described in (Smirnov et al., 1996; Smirnov et al.,

1994; Kiselev, Yumashev, 1990; Kiselev, 1994; Kiselev, 1997). The walls of containments can be of a multilayer composite structure thus the multiphase models are also preferable to describe the process and evaluate the accumulation of damages. The materials of the walls can be regarded as an elastoviscoplastic media (Smirnov *et al.*, 1996; Smirnov *et al.*, 1994) or thermoviscoelastic (Kiselev, 1997). The governing system of equations for the axisymmetrical problem of dynamical deforming of thermoviscoelastic composite shell (Kiselev, 1997; Kiselev, 1996) enables us to determine the growth of elastic energy and mechanical dissipation in the shell in dynamical deforming.

Generally internal loading of the walls of a tank in diffusive combustion of propellants is not uniform due to nonuniformities of combustion process and energy release caused by initial nonuniformities of concentrations of species. Theoretical models of dynamical internal loading of the shell show that the internal stresses in the shell depend upon loading of the shell by internal pressure $p(t)$. Nonuniformities of initial pressure distribution will cause the initial nonuniformities in shell stresses that will lead to dilatant wave propagation in the shell itself. The velocity of those waves will be

$$c_s = \sqrt{\frac{E}{(1-v^2)\rho_{s0}}}, \qquad (7.1)$$

where E, v are the Young modulus and Poisson coefficient of the material of the shell, ρ_{s0} — density of the material in the initial undisturbed state.

Stress wave propagation in the shell leads to creating a uniform stress field in the shell in a way similar to how compression and rarefaction waves in gas contribute to establishing a uniform pressure distribution inside the tank. The main difference is that characteristic wave propagation velocities in the shell c_s are of the order of magnitude higher than that in gas. Thus, the field of stresses in the shell will be uniform much earlier than that in the gas phase.

To perform some simple estimates one must examine the characteristic duration of the process of energy release inside a tank in diffusive combustion. Results of numerical modeling show that for a tank of characteristic size $r = 1$ m the characteristic time of combustion is:

$$\tau_c = 10^{-3} \div 10^{-2} \text{ s}.$$

The wave propagation velocity in aluminum alloy (7.1) is approximately 4.5 km/s.

Thus, the characteristic length of wave propagation during the combustion process for the model tank will be

$$L_s = 4.5 \div 45 \text{ m}$$

that is much larger than the characteristic size of the tank itself.

The estimates show that under these conditions the assumption of the uniform stress field established inside the shell by the end of combustion process is quite reasonable, while some nonuniformities in a pressure field inside the tank can be still present.

The characteristic times for wall loading and dynamical deforming of the walls are much less than that for combustion process: $\tau_e \approx 10^{-4}$ s. Thus dynamical deforming of the shell and breakup at early stages could be determined locally, while at later stages it should be regarded with account of averaged stresses.

The general statement of the problem of dynamical deforming and breakup of a shell and the governing equations are discussed in the Annex. A simplified statement of the problem, illustrating the dynamic effects is the following. The approach is based on a thin-walled shell theory. The following basic assumptions are made:

1. The shell is thin $h/r_* \ll 1$ (h — thickness, r_* — effective radius).
2. The material is considered to be elastoviscoplastic.
3. The field of tensile stresses is supposed to be uniform.
4. Momentum equation for a shell is solved locally with account of local internal loading, tensile stresses and spherical curvature with effective radius r_*.
5. The breakup conditions are considered to be determined on the basis of the entropic criterion of a limit specific dissipation.
6. The breakup process is supposed to take place as a result of the action of tensile ring stresses, consuming the elastic energy accumulated in the shell by the time of breakup.
7. The work of external forces in the breakup time, transformation of some part of elastic energy into the kinetic energy of fragments and spallation effects are not taken into consideration.

Then the momentum equation has the form:

$$\rho_s \dot{v} = \frac{p(t)}{h} - 2\frac{\sigma_\theta}{r}, \tag{7.2}$$

where ρ_s — density, v — radial velocity, r — current value of the shell radius of curvature, σ_θ — ring stress (the averaged stress over the shell thickness), a dot over a symbol means a material derivative with respect to time.

The rate of ring deformation is determined by:

$$\dot{\varepsilon}_\theta = \frac{v}{r}, \tag{7.3}$$

other deformations are absent due to the shell thinness.

The equation of mass conservation has the form $\dfrac{\dot{\rho}_s}{\rho_s} = -2\dot{\varepsilon}_\theta$ from where we have:

$$\rho_s = \rho_{s0} \exp(-2\varepsilon_\theta), \tag{7.4}$$

where ρ_0 is the initial density.

State equations of elastoviscoplastic material for a spherical case have the following form:

$$\dot{S}_\theta = \frac{2}{3}\hat{\mu}\dot{\varepsilon}_\theta - \frac{\hat{\mu}}{\eta}S_\theta\left(1 - \frac{J_0}{3|S_\theta|}\right) \times H\left(1 - \frac{J_0}{3|S_\theta|}\right), \tag{7.5}$$

where $\|S\|$ is a deviator of a stress tensor:

$$\sigma_\theta = \sigma + S_\theta, \quad \sigma_\varphi = \sigma_\theta, \quad \sigma_r = \sigma + S_r = 0, \quad 2S_\theta + S_r = 0, \quad \sigma_\theta = 3S_\theta. \tag{7.6}$$

Here $\hat{\mu}$ — shear modulus, η — dynamic viscosity of material, J_0 — static limit of elasticity in simple tension, $H(x)$ — the unit Heavyside function. Herewith the stress tensor σ_{ij} is divided into spherical $\sigma\delta_{ij}$ and deviator S_{ij} parts and it is assumed that rates of deformations can be divided into the rates of elastic and plastic ones and a plastic flow is incompressible:

$$\dot{\varepsilon}_\theta = \dot{\varepsilon}_\theta^e + \dot{\varepsilon}_\theta^p, \quad \dot{\varepsilon}_\varphi = \dot{\varepsilon}_\theta, \quad \dot{\varepsilon}_r = \dot{\varepsilon}_r^e + \dot{\varepsilon}_r^p = 0, \quad 2\dot{\varepsilon}_\theta^p + \dot{\varepsilon}_r^p = 0. \tag{7.7}$$

Specific (per mass unit) elastic energy E and mechanical dissipation D can be determined by integration of the formulae:

$$\dot{E} = 2\frac{\sigma_\theta}{\rho}\dot{\varepsilon}_\theta^e, \quad \dot{D}^0 = 2\frac{\sigma_\theta}{\rho}\dot{\varepsilon}_\theta^p. \tag{7.8}$$

Rates of elastic and plastic deformations are determined by the formulae:

$$\dot{\varepsilon}_\theta^e = \frac{\dot{S}_\theta}{2\mu} + \frac{2}{3}\dot{\varepsilon}_\theta, \quad \dot{\varepsilon}_\theta^p = \dot{\varepsilon}_\theta - \dot{\varepsilon}_\theta^e. \tag{7.9}$$

Calculation of the shell temperature in deforming for adiabatic approximation can be performed by the following procedure:

$$\rho c_\sigma \dot{T} + 2\alpha_v \dot{S}_\theta T = 6S_\theta \dot{\varepsilon}_\theta^p. \tag{7.10}$$

Herein c_σ is the heat capacity at constant stresses, α_v is the coefficient of volumetric expansion.

Criterion of the shell breakup is the entropic criterion of a limit specific dissipation that for the model considered is reduced to mechanical dissipation:

$$D^0\big|_{t=t_*} = D_*, \tag{7.11}$$

where D_* is the constant of limit specific dissipation determined with using the experiments on spalation fracture in a plane collision of plates.

The equation of state of material (7.5) contains static elasticity modulus in simple tension J_0 and shear modulus as parameters. It is known that in dynamical deforming of materials with essential changes of temperature, density, pressure and with growth of plastic deformations these parameters undergo changes. The elasticity limit J_0 and shear modulus μ are assumed to change in accordance with Steinberg-Guinane model (Kiseler, 1994):

$$J_0 = J_0^0 (1 + \beta \varepsilon_u^p)^n \left(1 - \beta \sigma \left(\frac{\rho_{s0}}{\rho_s} \right)^{\frac{1}{3}} - \chi (T_s - T_0) \right), \tag{7.12}$$

$$J_0^0 (1 + \beta \varepsilon_u^p)^n \leq J_{max}, $$

$$\mu = \mu_0 \left(1 - b \sigma \left(\frac{\rho_{s0}}{\rho_s} \right)^{\frac{1}{3}} - \chi (T_s - T_0) \right). \tag{7.13}$$

Here J_0^0, μ_0 are values of parameters under the normal conditions, $\varepsilon_u^p = 2 \left| \varepsilon_\theta^p \right|$ is the intensity of plastic deformations; β, n, b, χ, J_{max} are constants of the material.

The last term in the equation (7.12) shows that there is a dependence of elasticity limit on the mean wall temperature T_s. This temperature changes due to adiabatic heating of material in deformations and due to conductive heating of the wall contacting hot reaction products in combustion process.

The increase of temperature in dynamical deforming of the shell can be described by the equation (7.10).

The conductive heating of the wall depends upon the internal temperature of gases and the average expected time of heating. Results of numerical modeling of diffusive combustion inside tanks show that temperatures of gas near the walls can reach 3000 K in some places. That brings us to the necessity of estimating the wall heating in combustion process. The exact solution of the problem of heating of a wall being in contact with nonuniform varying with time gas flow is rather complicated. Thus, to obtain simple estimates we will use the following approach. The characteristic velocity of the thermal wave propagation in the material is

$$c_t = \frac{1}{2} \sqrt{\frac{\lambda_s}{\rho_s c_\sigma \tau}}, \tag{7.14}$$

where λ_s is a heat conductivity coefficient, τ — characteristic time of the process. Assuming that the material of the wall is characterized by the following parameters:

$$\lambda_s = 10^2 \ \text{w/m} \cdot \text{K}; \quad \rho_s = 2.7 \cdot 10^3 \ \text{kg/m}^3; \quad c_\sigma = 934 \ \text{J/kg} \cdot \text{K}$$

and characteristic time of the process $\tau_c = 10^{-3} \div 10^{-2}$ s, one obtains the values of velocities

$$c_t = 0.03 \div 0.09 \ \text{m/s}$$

The above estimates show that for the walls 2–3 mm thick the conductive heating during characteristic times $\tau_c \approx 10^{-3} \div 10^{-2}$ s will be negligibly small.

One can easily come to a conclusion that in case the destruction criterion is satisfied within the characteristic time of a rapid pressure increase in combustion, the conductive heating doesn't play any important role in the breakup process.

The situation can turn out to be different for the cases when the destruction criterion is not satisfied during the time of combustion. The shell is deformed and loaded by the end of combustion process and can stay in the position for a long time. Then the characteristic time of the whole process regarded can surpass the characteristic time of heat conductivity:

$$\tau_h = h^2 \bigg/ \left(\frac{\lambda_s}{\rho_s c_\sigma} \right) \tag{7.15}$$

By that time the increase of mean temperature of the wall can play an important role in the breakup process. The increase of temperature T_s will lead to the decrease of elasticity limit J_0 (equation (7.12)) that leads in turn to the decrease of tensile stress σ_θ, elastic strains ε_θ^e and causes increase of plastic strains ε_θ^p (equations (7.5), (7.6), (7.9)). Elastic energy accumulated by the shell will decrease and dissipation D^0 will increase (7.8).

Thus, the increase of temperature can lead to the situation when the increased dissipation D^0 surpasses the critical value D_* and the criterion of breakup is satisfied (7.11). Under these conditions the breakup process can take place much later after combustion and energy release processes are terminated. By the time pressure turns out to be uniform inside the tank and temperature uniformity is achieved due to intensive turbulent mixing of gases inside the tank and heat conductivity processes. The dynamics of mean wall temperature T variations in time can be determined from (7.10) by the following formula

$$T(t + \Delta t) = T(t) + \frac{1}{\rho_{s0} c_\sigma} (6 S_\theta \dot{\varepsilon}_\theta^p - 2\alpha_v \dot{S}_\theta T(t)) \Delta t +$$
$$+ \sqrt{\frac{\lambda_s}{\rho_{0s} c_\sigma}} \frac{T_{av} - T_0}{2h} (\sqrt{t + \Delta t} - \sqrt{t}) H(\tau_h - t) \qquad ; \tag{7.16}$$

where $H(x)$ is the unit Heavyside function; T_{av} — average temperature inside the tank near the wall. This temperature depending on characteristic time of the process can be determined either locally or with account of mixing inside the tank

$$T_{av} = \frac{\int_V \rho T dv}{\int_V \rho dv}.$$

Summing up the assumptions of the model it should be mentioned that the process under consideration can have three characteristic times:

- the characteristic time of wave propagation inside the shell itself $\tau_w = R^*/c_s$;
- the characteristic time of gasdynamics processes inside the tank τ_c;
- the characteristic time of heat conductivity τ_h.

The gasdynamics time is usually determined by a ratio

$$\tau_c = \frac{r^*}{a_c}, \tag{7.17}$$

where a_c is a sound velocity in gas. But within the frames of the problem regarded the gasdynamics processes are burdened with chemical processes and diffusive mixing. Thus, we determine the characteristic time τ_c not from the formula (7.17) but rather from the numerical solution of the problem of diffusive combustion of hypergolic propellants inside a fuel tank.

The following relationship between the characteristic times is valid:

$$\tau_w \ll \tau_c \ll \tau_h ; \tag{7.18}$$

$$\tau_w = \frac{r^*}{c_s} = \frac{r^*}{\sqrt{\dfrac{E}{(1-v^2)\rho_{s0}}}}; \quad \tau_c = \frac{r^*}{a_c}; \quad \tau_h = \frac{h^2 \rho_{s0} c_\sigma}{\lambda}$$

Since we are examining the breakup process the characteristic time for our problem τ_p will be the time interval between the ignition of mixture inside the tank and the breakup of the whole tank.

If $\tau_p \sim \tau_w$, then nonuniformity of parameters both in the shell and in the gas phase is essential, heat conductivity does not play any role.

If $\tau_p \sim \tau_c$, then only gas phase nonuniformities are essential, and nonuniformities in the shell and heat conductivity do not play any role.

If $\tau_p \sim \tau_h$ then nonuniformities of stresses in gas and shell are not important but heat conductivity can play an important role.

Analyzing the problem we found out that

$$\tau_p > \tau_c .$$

Thus, it is necessary to foresee several possibilities within the frames of numerical modeling: breakup conditions can be reached within the time of combustion ($\tau_p \sim \tau_c$), the criterion of destruction can be satisfied after considerable heating of the shell ($\tau_p \sim \tau_h$), or the breakup criterion can be never satisfied ($\tau_p \to \infty$).

7.2.3. Fragmentation Models for Thin-walled Containments

We used the thermodynamic criterion of destruction based on the critical value of the dissipation in irreversible transformations: viscous dissipation, accumulation of damages in tension, shear and delamination. As soon as the destruction criterion is satisfied: $D \geq D_*$, a breakup of a shell takes place (Smirnov et al., 1996; Smirnov et al., 1994; Kiselev, 1996).

The scenario of fragmentation for small characteristic times of the process τ_p can differ from that for the case of uniform loading. The breakup can start in the zones of maximal internal loading and then cracks will develop in the zones of lower internal loading but with the same level of tensile ring stresses. The first stage of the process within the zones of high internal loading is initiated by the increase of internal pressure and growth of dissipation D.

In the zones of lower internal loading the rate of loading is not enough to initiate the breakup process. Thus in the second stage the breakup process is initiated by the cracks coming from the damaged zone.

The common point of the both scenario is that in both cases the destruction is governed by the balance of the accumulated elastic energy and the work consumed to break the material (to form the cracks).

The fragmentation of a tank will be regarded locally with account of nonuniformity of internal loading resulting in nonuniformity of elastic energy distribution inside the shell.

The number of fragments obtained in a shell breakup can be found from the balance of the elastic energy accumulated by the shell by the time of breakup and work necessary for the cracks formation. The elastic energy accumulated by the time of breakup t_* in an arbitrary section of the shell with an area S_α and thickness h can be determined by the formula:

$$E = \iint_{S_\alpha} E_* \rho_0 h dS \tag{7.19}$$

This energy is spent to form cracks (free surfaces) around fragments:

$$E \cdot k_E = N_{0\alpha} \int_0^\infty \gamma h p(s) \, f(s) ds \tag{7.20}$$

where γ is the specific energy consumed for formation of a free surface unit, $p(s)$ is the semiperimeter of a fragment of area s, $k_E (0 < k_E \leq 1)$ is the coefficient of the elastic energy consumption, $f(s)$ — the density of fragments distribution, $N_{0\alpha}$ — the number of fragments formed in breakup of the α-section of the shell.

To describe the probability distribution function for fragments versus area a modified Weibull distribution is used. The fragments mass in breakup of a thin shell is proportional to the area: $m = \rho_0 hs$. Therefore the following distribution could be used

$$N(<s) = N_0 F(s) \tag{7.21}$$

where

$$F(s) = \begin{cases} 1 - \exp\left(-\left(\dfrac{s - s_{min}}{s_*}\right)^{\Lambda}\right), & s_{min} < s < s_{max} \\ 0, & s < s_{min} \\ 1, & s > s_{max} \end{cases}$$

(7.22)

Formula (7.22) involves the following notations:

s_{min} — minimal possible area of fragment, s_{max} — maximal possible area of fragment.

Both values are to be determined in the process of solution.

Parameters Λ and s_* characterize the distribution function F(s) and could be determined from an independent experiment. We shall use the values of these parameters determined in (Smirnov et al., 1997) for breakup of thin shells in uniform loading (Fucke, 1993).

The density of distribution of fragments versus area can be obtained from formulas (7.21), (7.22):

$$N(ds) = N_{0\alpha} f(s) ds = N_{0\alpha} F'(s) ds \; ;$$

(7.23)

$$f(s) = \begin{cases} 0, & s < s_{min}; \\ \dfrac{\Lambda}{s - s_{min}}\left(\dfrac{s - s_{min}}{s_*}\right)^{\Lambda} \exp\left(-\left(\dfrac{s - s_{min}}{s_*}\right)^{\Lambda}\right), & s_{min} < s < s_{max}; \\ 0, & s > s_{max}. \end{cases}$$

The total area of fragments must be equal to the initial area of section S_α

$$S_\alpha = \int_0^\infty s N(ds) = N_{0\alpha} \int_0^\infty s f(s) ds.$$

(7.24)

Taking into account the type of the function $f(s)$ (7.23) one can obtain the following formula

$$S_\alpha = N_{0\alpha} \int_{s_{min}}^{s_{max}} s f(s) ds.$$

(7.25)

The function $f(s)$ has singularity at $s = s_{min}$ for the values $\Lambda < 1$. But the integral is finite.

Thus to determine $N_{0\alpha}$ one obtains formula, that follows from (7.25):

$$N_{0\alpha} = S_\alpha \left(\int_{s_{\min}}^{s_{\max}} sf(s)\,ds \right)^{-1}. \tag{7.26}$$

To determine $s_{\max}$ one must take into account that the maximal area of fragments can't exceed the total area of the section S_α. Formula (7.21) is valid under the assumption $N_0 > 1$. Thus following the assumption

$$N(<s) < N_0 : s_{\min} < s < s_{\max}$$

one can determine $s_{\max}$ by formula

$$N(<s_{\max}) = N_0 - 1 ,$$

that gives the following estimate

$$1 = N_{0\alpha} \exp\left(-\left(\frac{s_{\max} - s_{\min}}{s_*} \right)^\Lambda \right) \tag{7.27}$$

or

$$s_{\max} = s_{\min} + s_* (\ln N_0)^{1/\Lambda}. \tag{7.28}$$

Equations (7.26) and (7.28) give us the possibility to determine the values $N_{0\alpha}$ and $s_{\max\alpha}$ for the section regarded.

Equation (7.20) can be rewritten with account of (7.24) in the following form

$$N_{0\alpha} \int_0^\infty \left(p(s) - \frac{\rho_{0s} E_* \alpha k_E}{\gamma} s \right) f(s)\,ds = 0, \tag{7.29}$$

or with account of real density of distribution (7.23):

$$N_{0\alpha} \int_{s_{\min}}^{s_{\max}} \left(p - \frac{\rho_{0s} E_* \alpha k_E}{\gamma} s \right) f(s)\,ds = 0. \tag{7.30}$$

The relative arbitrariness of the limits of the integral (7.30) drives one to a conclusion that the equation is satisfied when the function under integral equals to zero for all types of fragments:

$$p = \frac{\rho_{0s} E_* \alpha k_E}{\gamma} s. \tag{7.31}$$

Physically the equation (7.31) means that one half of the energy necessary to form a breakup surface around each fragment with area s is extracted from the mass of the fragment itself and the other half — from outside (from the neighboring fragments).

On introducing a dimensionless coefficient of shape $k = s/p^2$ and keeping in mind that the shape coefficient can change within the limits $0 < k \le 1/\pi$ one can obtain some restrictions for the smallest possible area of fragments within the regarded section

$$s_{\min}^{\alpha} = \pi \left(\frac{\gamma}{\rho_0 E_{\bullet\alpha} k_E} \right)^2. \tag{7.32}$$

Equations (7.26), (7.28) and (7.32) form a closed system to determine the number and distribution of fragments within each section.

In case the amount of elastic energy is too low and the restriction (7.32) prohibits formation of fragments

$$\pi \left(\frac{\gamma}{\rho_{0s} E_{\bullet\alpha 1} k_E} \right)^2 > S_{\alpha 1}, \tag{7.33}$$

then two neighboring sections can be reunified to form a section $S_{\alpha 12} = S_{\alpha 1} + S_{\alpha 2}$ and the restriction (7.32) is checked once more for the mean value of elastic energy $E_{\bullet\alpha 12}$.

To perform the calculations one should extract K groups of fragments with definite areas s_j $(j = 1,\ldots,K)$ from the total spectrum of the shell fragments. Then the number of fragments within each group can be determined by a formula:

$$N_{j\alpha} = \left[F_\alpha \left(\frac{s_j + s_{j+1}}{2} \right) - F_\alpha \left(\frac{s_{j-1} + s_j}{2} \right) \right] N_{0\alpha}.$$

The total number of fragments of the j-th group obtained in a breakup of the whole vessel is determined by summing up all fragments of the size in all sections of the shell:

$$N_j = \sum_{\alpha=1}^{L} N_{j\alpha}.$$

Final velocities of fragments can be determined solving the following equations:

$$m_j \frac{d\vec{v}_j}{dt} = -hs_j \frac{\partial p}{\partial r}\bigg|_{r=x} - \frac{c_d}{2} S\rho |\vec{v}_j - \vec{u}| (\vec{v}_j - \vec{u}),$$

$$\frac{dx_j}{dt} = \vec{v}_j,$$

where

$$S = s_j \cos\alpha + \left(h + \frac{l_j^2}{4r_0} \right) l_j \sin\alpha; \quad c_d = c_d(\alpha, f_j), \quad f_j = \frac{l_j^2}{s_j}, \quad m_j = \rho_s hs_j,$$

m_j is the mass of a fragment, $\vec{v}_j$ — fragment's velocity, $\vec{u}$ — velocity of gas, s_j — area of the surface (one side), h — fragment's thickness, c_d — the drag coefficient, S — the effective area facing the flow, α is the angle of the orientation, l_j is the characteristic size of a fragment, f is the shape coefficient, r_0 — the curvature radius of fragments. The initial conditions for the equation are the following:

$$t = 0 : \vec{v}_j = \vec{v}_{j\bullet}, \quad x = r_{0j},$$

where $\vec{v}_{j\bullet}$ is the velocity of the shell just before the breakup.

7.2.4. Fragmentations in Collisions of Debris Particles

The present section contains a description of a simple model for fragmentation in a hypervelocity collision of two particles, making it possible to evaluate number, mass and velocity distributions of fragments.

a) Mean velocities and energies of fragments

Let M_1 and M_2 be the masses of two particles before the collision; $\vec{V}_1$, $\vec{V}_2$ — their velocities. Then the velocity of the center of masses of the cloud of fragments after the collision will be the following:

$$\vec{V} = \frac{M_1 \vec{V}_1 + M_2 \vec{V}_2}{M}; \; M = M_1 + M_2 \qquad (7.34)$$

In collisions some part of the kinetic energy of particles is transformed into the internal energy

$$\frac{M_1 \vec{V}_1^2}{2} + \frac{M_2 \vec{V}_2^2}{2} = \frac{MV^2}{2} + U; \; U = U_1 + U_2, \qquad (7.35)$$

where the internal energy U is a sum of elastic and non-elastic energies (dissipations):

$$U_1 = E_1 + D_1; \; U_2 = E_2 + D_2$$

Intensive characteristics: densities of energy and dissipation — could be defined for the particles.

$$U = Mu; \; U_1 = M_1 u; \; U_2 = M_2 u; \; E_1 = M_1 e_1;$$
$$E_2 = M_2 e_2; \; D_1 = M_1 d_1; \; D_2 = M_2 d_2$$

Fragmentation taking place in collisions is a result of complex thermomechanical wave processes of irreversible deforming and microfracture. It is assumed that the criterion for the macrofracture of material is that of the critical specific dissipation (section 7.2.2). Then one could assume that by the time of breakup the folllowing equalities are valid:

$$d_1 = d_1^{\bullet}; \; d_2 = d_2^{\bullet},$$

where d_1^*, d_2^* are the known material parameters that could be determined independently in experiments on spallation plates formation (Kiselev, Yumashev, 1990; Kiselev, 1994). The elastic energies of both particles after the collision could be:

$$E_1 = \frac{M_1}{M} U - M_1 d_1^*; \quad E_2 = \frac{M_2}{M} U - M_2 d_2^*$$

Some part of this energy will be used for breakup (cracks formation):

$$E_1^f = k_E E_1, \quad E_2^f = k_E E_2, \quad (0 \le k_E \le 1)$$

For $k_E = 1$ all the elastic energy is spent for fragmentation. If the internal energy accumulated by a particle in collision is less than the critical dissipation $U_\alpha \le M_\alpha d_\alpha^*$, $(\alpha = 1,2)$, then the fragmentation does not take place.

b) Number of fragments

The number of fragments is determined using the same approach we used in the models for thin-walled structures fragmentations (section 7.2.3) based on the Weibul distribution:

$$N(<m) = N_0 \left[1 - \exp\left(-\left(\frac{m - m_{\min}}{m_*} \right)^\Lambda \right) \right], \quad m_{\min} \le m \le m_{\max} \qquad (7.36)$$

One could distinguish K_α groups of fragments for each particle $(\alpha = 1,2)$ characterized by different masses $m_1^\alpha, m_2^\alpha, ..., m_{K_\alpha}^\alpha : m_{\min}^\alpha < m_1^\alpha < ... < m_{K_\alpha}^\alpha < m_{\max}^\alpha$, where $m_{\min}^\alpha, m_{\max}^\alpha$ — minimal and maximal sizes of fragments. The distinguished groups are assumed to incorporate the following fragments:

$$m_1^\alpha : \sqrt{m_{\min}^\alpha m_1^\alpha} \le m^\alpha \le \sqrt{m_1^\alpha m_2^\alpha};$$
$$m_2^\alpha : \sqrt{m_1^\alpha m_2^\alpha} \le m^\alpha \le \sqrt{m_2^\alpha m_3^\alpha};$$
$$...$$
$$m_{K_\alpha}^\alpha : \sqrt{m_{K_{\alpha-1}}^\alpha m_{K_\alpha}^\alpha} \le m^\alpha \le \sqrt{m_{K_\alpha}^\alpha m_{\max}^\alpha}.$$

Then the number of fragments of the j-th group $(j = 1, 2, ..., K_\alpha)$ is the following:

$$N_j^\alpha = N_0^\alpha (\beta_j^\alpha - \beta_{j+1}^\alpha), \quad \beta_j^\alpha = \exp\left[-\left(\frac{\sqrt{m_{j-1}^\alpha m_j^\alpha} - m_{\min}^\alpha}{m_*^\alpha} \right)^{\Lambda_\alpha} \right]. \qquad (7.37)$$

where N_0^α is the total number of fragments generated in breakup of the α-th particle.

The system (7.37) contains K_α equations. Equations of mass and energy conservation in fragmentation could be added to the system (7.37):

$$\sum_{j=1}^{K_\alpha} m_j^\alpha N_j^\alpha = M_\alpha; \quad \alpha = 1,2 \tag{7.38}$$

$$\sum_{j=1}^{K_\alpha} \gamma_\alpha \frac{s_j^\alpha}{2} N_j^\alpha = E_\alpha^f, \quad \alpha = 1,2 \tag{7.39}$$

where γ_α — specific energy for free surface formation; s_j^α — free surface area for the m_j^α fragments; $E_\alpha^f = e_\alpha^f M_\alpha$; e_α^f — specific energy spent for fragmentation. The equations (7.38), (7.39) make it possible to obtain the following:

$$\sum_{j=1}^{K_\alpha} \left(\gamma_\alpha \frac{s_j^\alpha}{2} - m_j^\alpha e_\alpha^f \right) N_j^\alpha = 0, \quad \alpha = 1,2 \tag{7.40}$$

Thus to determine $[(2K_\alpha + 1)\cdot 2]$ unknown parameters: s_j^α, N_j^α, N_0^α one has only $2(K_\alpha + 2)$ equations (7.37)–(7.39). A particular solution of the equation (7.40) could be added to the system of equations

$$\gamma_\alpha \frac{s_j^\alpha}{2} = m_j^\alpha e_\alpha^f, \quad j = 1,2,\ldots,K_\alpha; \quad \alpha = 1,2 \tag{7.41}$$

The physical meaning implied in the equations (7.41) is the following: the right hand part of (7.41) is the elastic energy of the m_j^α fragment spent for fragmentation. The left hand part of (7.41) is a half of the energy needed to form the fragment. The other half of the necessary energy is taken from the neighboring fragments.

The equations (7.37), (7.38) give the following formulas:

$$N_0^\alpha = \frac{M_\alpha}{\displaystyle\sum_{j=1}^{K_\alpha} m_j^\alpha (\beta_j^\alpha - \beta_{j+1}^\alpha)}; \quad N_j^\alpha = N_0^\alpha (\beta_j^\alpha - \beta_{j+1}^\alpha), \quad j = 1,2,\ldots,K_\alpha. \tag{7.42}$$

The equations (7.41), (7.42) form a closed system of $(4K_\alpha + 2)$ equations to determine the unknown parameters.

c) *The minimal mass of fragments*

The equations (7.41) provide a link between the mass of a fragment m_j^α and the area of cracks S_j^α being a part of the fragments free surface area $\left(S_j^\alpha \dfrac{s_j^\alpha}{S_j^\alpha} = \theta_j^\alpha; \ 0 < \theta_j^\alpha \le 1 \right)$.

Introducing fragments volumes $A_j^\alpha = \dfrac{m_j^\alpha}{\rho_\alpha}$ the equation (7.41) illustrates the dependence of the volume of fragments and the fragmentation area.

On introducing shape coefficients for fragments $k_f = \dfrac{A}{\sqrt{S^\alpha}}$ and taking into account that

$$0 \le k_f \le \frac{1}{6\sqrt{\pi}},$$

one obtains the following estimate

$$m_j^\alpha \ge \frac{9\pi\theta_j^{\alpha^2}}{2\rho_\alpha^2}\left(\frac{\gamma_\alpha}{k_E e_\alpha}\right)^3 \tag{7.43}$$

The restriction (7.43) should be taken into account while choosing the groups of fragments. The mass of minimal fragment:

$$m_{\min}^\alpha = \frac{9\pi}{2\rho_\alpha^2}\left(\frac{\gamma_\alpha}{k_E e_\alpha}\right)^3 \tag{7.44}$$

The equation (7.44) shows that the increase of the elastic energy accumulated by the time of breakup increases the fragmentation intensity (formation of smaller fragments).

d) Relative velocities of fragments

Assuming the rest of elastic energy of fragments $(1 - k_E)E_\alpha$, $(\alpha = 1,2)$ to be transformed into the kinetic energy of the fragments scattering in the co-ordinate system connected with the center of masses of the system, one obtains the following relationship:

$$(1-k_E)E_\alpha = \sum_{j=1}^{K_\alpha} \frac{m_j^\alpha (v_j^\alpha)^2}{2} N_j^\alpha \tag{7.45}$$

Keeping in mind that $E_\alpha = e_\alpha M_\alpha = e_\alpha \sum_{j=1}^{K_\alpha} m_j^\alpha N_j^\alpha$, one could transform (7.45) in the following way:

$$\sum_{j=1}^{K_\alpha}\left(\frac{(v_j^\alpha)^2}{2} - (1-k_E)e_\alpha\right)m_j^\alpha N_j^\alpha = 0 \tag{7.46}$$

The last equation has the following particular solution:

$$v_j^\alpha = \sqrt{2(1-k_E)e_\alpha}, \tag{7.47}$$

characterized by equal radial velocity values for all the fragments.

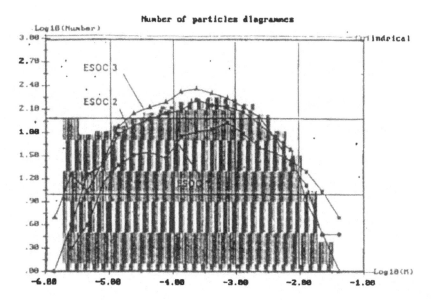

Fig. 7.8. Number of fragments versus mass distribution for the breakup of a cylindrical vessel $r_0 = 30$ cm with wall thickness $h = 0.5$ mm, and density 27(X) kg/m^3, in uniform loading by a detonation wave in hydrogen-oxygen stoicheometric mixture. Theoretical results are represented by the diagram; experimental results are illustrated by the curves.

The developed mathematical model for fragmentation of debris particles in collisions could be incorporated as one of the source terms into the models developing long-term forecasts for space debris evolution. The developed model is the first one, to our knowledge, model of fragmentation based on thermodynamic breakup criterion of the critical dissipation in irreversible transformations.

7.3. Fragmentation of Shells in Uniform Internal Loading

A breakup of a cylindrical thin walled containment was regarded caused by an explosion of hydrogen-oxygen mixture filling the containment. The parameters of the mixture and the containment were chosen corresponding to that for the experiments described in (Fucke, 1993).

The Fig. 7.8 shows one of the test cases for the fragmentation event in the form of a diagram of total number of fragments of different masses originating as a result of a breakup of a cylindric shell caused by detonation of hydrogen-oxygen mixture inside.

The calculated diagram corresponds to experiment ESOC-2 (Fucke, 1993). The initial conditions were the following: the tank radius $r = 30$ cm, length 75 cm, wall thickness $h = 0.5$ mm, density $\rho = 2700$ kg/m^3, heat capacity $c_v = 924.3$ J/kg·K, volume extendibility $\alpha_v = 6.72 \cdot 10^{-5}$ K^{-1}, shear module $\mu = 27$ GPa, elasticity limit $J_{max} = 0.68$ GPa, energy per breach square unit $\gamma = 100$ kJ/m^2, exponential parameter $\Lambda = 0.5$. Comparison of the theoretical and experimental results shows that for the part of spectrum containing large fragments the coincidence is rather fine. A slight difference for small fragments (theory gives a larger number than the experiment) can be explained by the fact that in the experiment not all the fragments were collected. It is mentioned in (Fucke, 1993) that for the test case ESOC-2 the lost mass was about 1.3% of the total mass of the shell that equals to about 20g. The existing difference of mass for small fragments between theoretical and experimental results is much less than this value.

Different modes of energy release processes were regarded: deflagration, detonation and deflagration to detonation transition process (Smirnov, Panfilov, 1995; Smirnov, Tyurnikov, 1994). In the last case an overdriven detonation wave can appear, producing very high rates of loading in reflection (Smirnov *et al.*, 1997).

The results of numerical modeling show that the breakup process, the number and mass distribution of fragments, and their velocities differ greatly depending on the combustion mode (rate of energy release) inside the tank and this dependence is not monotonous. For the lowest (deflagration) and highest (detonation) rates of energy release the number of fragments is less than for the medium rate of energy release: i.e. deflagration to detonation transition. Maximal fragments' velocities and the time before the breakup for those three cases are illustrated by the following table.

Parameter	Deflagration	Transition process	Detonation
breakup time (ms)	7017	4272	1029
maximal velocity (m/s)	900	2500	1000

This dependence of breakup characteristics on the rate of the energy release inside the fuel tank is too large to be neglected.

The next series of numerical experiments was devoted to testing the so-called similarity parameters for breakup of fuel tanks (Fucke, 1993). The similarity parameter was supposed to be $p_0 r_0/h = A = $ const. It was supposed in (Fucke, 1993) that for the constant values of similarity parameters the number of fragments and their final velocities remained constant and masses of large fragments grew up proportionally to the growth of the volume of the shell's material.

The calculations were performed for the case similar to ESOC-2 but with proportionally increased radial dimensions of the shell and thickness: $r_0 = 120$ cm, $h = 2$ mm. This increase of dimensions of the tank preserves the similarity parameter A constant. The results showed that the number of fragments increased nearly in an order of magnitude contrary to predictions (Fucke, 1993). The increase of final velocity was more than 20%. The volume of the shell material increased 16 times, but the mass of largest fragments contrary increased only 8–9 times.

Thus the results show that there cannot exist similarity parameters for such a complicated phenomena as breakup of a fuel tank and scaling of model tanks is a very difficult procedure. To apply the results of model experiments to the real fuel tank it is necessary to make use of general theoretical models and not simple upscaling parameters.

7.4. Breakups Caused by Non-uniform Internal Loading

Most of explosions in space are not that producing uniform loading on the construction elements. Some of explosions were caused by accidental mixing of hypergolic components in fuel tanks of the second stages that had been in orbit for significant periods of time (Loftus, 1987; NASA-JSC Technical Memorandum, 1981; Webster, Kawamura, 1992). The conditions arose because first stage systems shut down on fuel or oxidizer depletion. Performance reserves are all in the second stage. The failure of the bulkhead brings to mixing and ignition of propellants. The energy release in breakup is not consistently related to residual propellant mass but to the mass of mixture of self-igniting components in an unknown pattern of dispersal within the tank formed after the breakup of the bulkhead. This type of combustion studied in (Smirnov et al., 1996) brings to nonuniform and sometimes oscillating loadings of the structure.

The fragmentation scenario in nonuniform loading can differ from that in uniform loadings. The breakup can start in the zones of maximal loadings where the breakup criterion is satisfied, and then cracks can develop in less loaded zones. This second stage of breakup can be initiated by the cracks coming from the damaged zone. Thus the less loaded zones (wherein the accumulated elastic energy E_{α^*} is smaller) form the larger fragments in breakup. Nonuniformities in internal loading can lead to different spectra of mass distribution of fragments.

Numerical investigations of wall-loading and breakup in diffusive combustion inside the tanks were carried out for the cylindrical tank of 1 m diameter and 1 m length with the walls 2 mm thick and a coaxial orifice in the bulkhead 25 cm radius (Smirnov et al., 1996).

On opening the orifice hypergolic components came in contact and a diffusion combustion process started. The dynamics of the process in a two-chambers containment was discussed in the section 7.2.1. The pressure growth pattern generated by the unsteady combustion process is very irregular. Figure 7.9a shows the wall pressure profile formed by the time 2.12 ms after the ignition. The coordinate axis "s" starts in the center of the bottom part of the cylinder and following the wall reaches the center of the top plate (*ABCD* in Fig. 7.5). Vertical lines in Fig. 7.9 mark the connections of the side wall of the cylinder with the top and bottom plates. The tangential stresses in the walls at the same time are shown in Fig. 7.9b.

Figures 7.10 and 7.11 show the plots of dissipation D and elastic energy E of the shell in a logarithmic scale for the time corresponding to that in Fig. 7.9. The breakup took place when dissipation in one of the sections overcame the critical value D^* determined in independent experiments on spallation. The total number of fragments versus mass distribution function is shown in Fig. 7.12. The corresponding velocities distributions are shown in Fig. 7.13 in the form of the plots for maximal, mean and minimal velocities.

It is seen from Fig. 7.12 that the plot of number of fragments versus mass distribution has two maxima and one of them is for the large fragments, that is an essential difference from the distribution (Fig. 7.8) obtained for symmetrical loading by detonation. Large fragments were formed in the less loaded zones and small fragments were formed in the zones of high density of the accumulated elastic energy E_*.

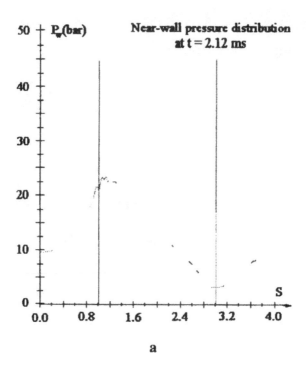

a

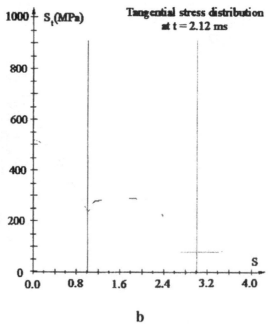

b

Fig. 7.9. The wall pressure (a) and tensile stresses (b) distribution in the shell of the two-chamber vessel loaded by internal diffusion combustion. $t = 2.12$ ms after the ignition.

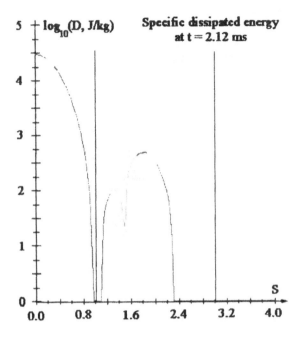

Fig. 7.10. Specific dissipation accumulated by the shell in non-uniform internal loading: $t = 2.12$ ms after the ignition inside the tank.

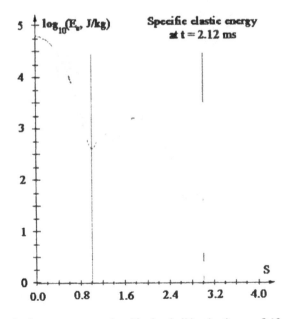

Fig. 7.11. Specific elastic energy accumulated in the shell by the time $t = 2.12$ ms after the ignition inside the tank.

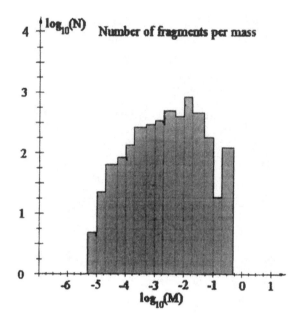

Fig. 7.12. The total number of fragments versus mass distribution in breakup of the non-uniformly loaded shell.

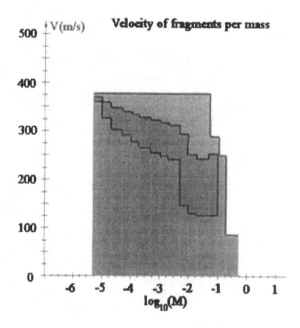

Fig. 7.13. The maximal, mean and minimal velocities of fragments versus mass distribution in breakup of a nonuniformly loaded shell.

Cumulative number of fragments per square root of mass

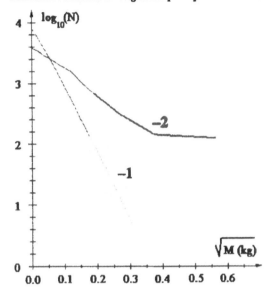

Fig. 7.14. The cumulative flux of fragments generated in breakup of a shell 1 m in diameter, 1 m length with the walls 2 mm thick and a coaxial orifice in the bulkhead 50 cm diameter. Results of numerical modeling. Curve 1 — uniform loading by a detonation wave in a premixed hydrogen-oxygen mixture. Curve 2 — nonuniform loading due to diffusive combustion of non-premixed hypergolic propellants.

The fragments velocity distributions (Fig. 7.13) show that the mean velocity for large fragments is less than that for the smaller ones. Final velocity for the very small fragments does not depend on mass and is practically constant that is in a good agreement with the velocity distributions assumed by (Chobotov, Spencer, 1991). The decrease of velocity for large fragments took place due to the fact that in nonuniform loading the less loaded zones accumulated less elastic energy and less kinetic energy. Those zones could produce only large fragments in breakup.

The cumulative flux of fragments for the regarded case is shown in Fig. 7.14 (curve 2). For comparative purposes Fig. 7.14 contains also the cumulative flux of fragments formed in breakup of the same vessel in uniform loading by the detonation of hydrogen-oxygen mixture (curve 1). The difference in cumulative fluxes for the explosions of one and the same energetic equivalent is due to different modes of energy release and different loading patterns. Fig. 7.15 contains the comparison of the cumulative number of fragments detected in orbital breakup of the Himawari sattelite with the EVOLVE model predictions (Potter, Anz-Meador *et al.*, 1994). The qualitative comparison with Fig. 7.14 shows that the breakup having taken place was rather one caused by nonuniform loading in combustion than a high intensity explosion. While the EVOLVE model predictions were based on an assumption of high intensity explosions to be the cause of obital breakups.

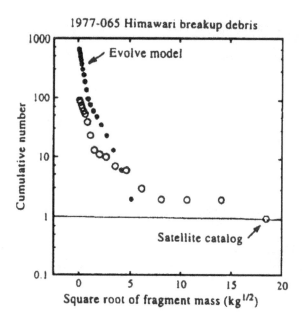

Fig. 7.15. The cumulative flux of fragments detected in orbital breakup of Himawari (circles) and Evolve predictions (dots) (Anz-Meador, Potter, 1996).

7.5. Fragmentations Caused by Hypervelocity Collisions of Debris Particles with Pressurized Vessels

Fragmentation of a gas-filled or fluid-filled containment in hypervelocity collision has several characteristic stages. The first stage is fragmentation of the impactor and the wall in the collision zone and formation of a hypervelocity jet of small fragments penetrating inside the containment. Formation of cracks (and petals) in the collision zone does not usually bring to a breakup of the containment at the present stage. The hypervelocity fragments cloud forms a shock wave in the media, filling the containment. In case of a highly compressible media (gas) the edges of the hole in the wall (or petals) are deformed inside the containment, in case of a low compressibility of the media (fluid) the edges of the hole are bulging out.

The cloud of small fragments slows down very rapidly due to the drag forces. The deceleration for fragments is proportional to $1/r_0$ and grows up with the decrease of a characteristic size r_0. On slowing down the cloud the convertion of its kinetic energy into the internal energy of the surrounding gas (or fluid) takes place. The rapid increase of the density of energy in a small volume inside the containment is similar to that for the local explosion. The energy release gives birth to diverging blast waves inside the containment that reflect from the walls thus producing nonuniform loading. The loading pattern is somehow similar to that in reflected detonation waves but the fragmentation scenario can be different for the same energy equivalent. The concentrated energy release

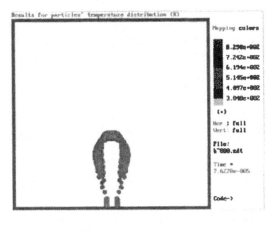

a

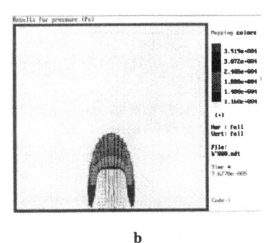

b

Fig. 7.16. Model fragments location (a) and gas pressure distribution (b) inside the gas-filled containment ($p_0 = 0.01$ MPa) at a time $t = 76\ \mu s$ after a hypervelocity impact.

causes blast waves of higher intensity than the detonation wave. Thus the wall being more close to the blast point exercises higher loading and gives birth to a large number of small fragments. The breakup of the wall causes the pressure drop and the rarefaction waves going inside the containment, overtaking the blast wave and lowering down its intensity. This is not the case for the detonation wave propagating inside the containment as the rarefaction waves from the depressurization zones will never overtake the detonation wave and never decrease its intensity. Hence the "point blast" energy release could produce a more wide spectrum of fragments than the detonation wave loading.

Figures 7.16–7.21 present the results of mathematical modeling of hypervelocity cloud of fragments propagation in a gas-filled cylindrical containment after perforation of the

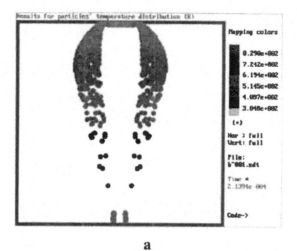

a

b

Fig. 7.17. Model fragments location (a) and gas pressure distribution (b) inside the gas-filled containment ($p_0 = 0.01$ MPa) at a time $t = 214$ μs after a hypervelocity impact.

bottom wall near the axis being the result of a hypervelocity impact. A cylindrical containment of 0.1 m radius and 0.2 m height was regarded, wall thickness 2 mm. A hole 10 mm in diameter was formed as a result of the impact, and the material of the wall formed a cloud of small fragments, characterized by the average diameter 0.3 mm and stochastic deviations 0.05 mm. The fragments initial temperature was 700 K with stochastic deviations ±50 K; maximal velocity in the axial direction was assumed to be 1900 m/s, average velocity 1500 m/s with stochastic deviations 400 m/s both in axial and radial directions. The average density of the material was assumed as $\rho = 2000$ kg/m^3, the melting temperature 800 K, viscosity and surface tension in the liquid state 10^{-3} Ns and

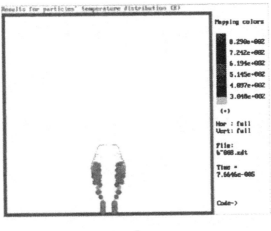

a

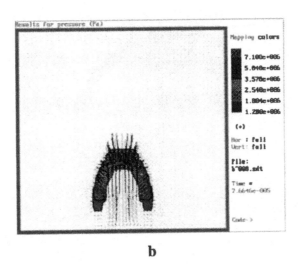

b

Fig. 7.18. Model fragments location (a) and gas pressure distribution (b) inside the gas-filled containment ($p_0 = 1$ MPa) at a time $t = 76$ μs after a hypervelocity impact.

10^{-2} N/m respectively. The gas pressure inside the containment was varied from 0.01 MPa up to 1.5 MPa, initial temperature $T_0 = 300$ K, molar mass 0.028 kg/mol.

The primary goal of the modeling was to follow the transformation of the kinetic energy of the fragments cloud and its contribution to internal loading of the containment. The secondary goal was to investigate if the internal atmosphere filling the containment, could serve as an additional bumper shield to protect the upper wall from perforation.

Figures 7.16, 7.17 show the model particles location (a) and pressure distribution inside the containment (b) for the two successive times. The initial pressure of gas inside the containment was rather low: $p_0 = 0.01$ MPa (0.1 bar). The line segments in Figs. 7.16b,

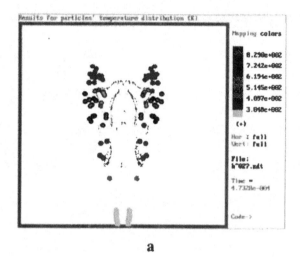

a

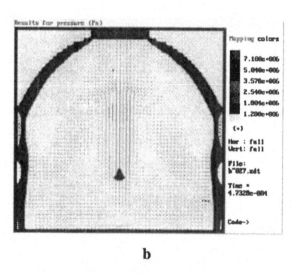

b

Fig. 7.19. Model fragments location (a) and gas pressure distribution (b) inside the gas-filled containment ($p_0 = 1$ MPa) at a time $t = 473$ μs after a hypervelocity impact.

7.17b illustrate directions and values of velocity vectors in gas. The size of circles showing model particles is much larger than their real size, but directly proportional to it.

Figures 7.16, 7.17 show that the cloud of fragments generates compression waves in the gas, but the velocity of the axial propagation of the cloud is too high for the shock waves to overtake it.

The shock wave and the cloud both collide the upper wall of the containment practically simultaneously in the present case of rather low initial pressure of gas filling the containment. Nevertheless, due to the dispersion of fragments in the cloud the total momentum is distributed on a larger area of the upper wall in collision.

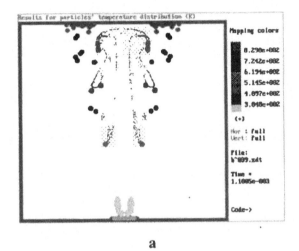

a

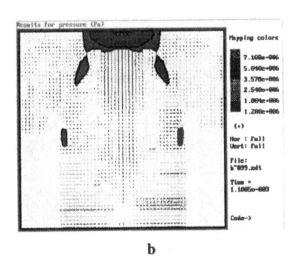

b

Fig. 7.20. Model fragments location (a) and gas pressure distribution (b) inside the gas-filled containment ($p_0 = 1$ MPa) at a time $t = 1100\,\mu$s after a hypervelocity impact.

The average pressure on the walls of the containment (taking into account the momentum of fragments in impact) is illustrated in Fig. 7.21a (curve 0.1 bar). The maximal loading still takes place at the axis of symmetry.

Figures 7.18, 7.19, 7.20 illustrate the model particles locations (a) and gas pressures (b) for the case of a relatively high gas pressure inside the containment ($p_0 = 1$ MPa). The aerodynamic drag and heating of the particles are much more essential for the present case. On entering the containment the front particles of the cloud are heated above the melting temperature. Fragmentation of liquid droplets due to their interaction with the atmosphere brings to a formation of very fine droplets in the front part of the cloud

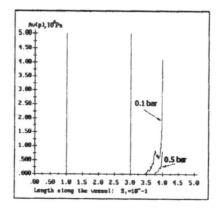

a

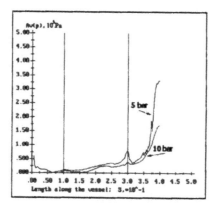

b

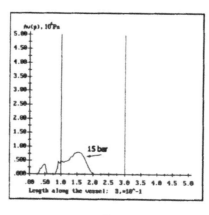

c

Fig. 7.21. Average wall overpressure distribution accounting for the momentum of impacting fragments for different initial gas pressures: 0.1 bar, 0.5 bar, 5 bar, 10 bar, 15 bar.

(Fig. 7.18a). A strong shock wave is formed ahead of the cloud (Fig. 7.18b). The particles, representing smaller fragments, are illustrated by dots in Fig. 7.18a. Nevertheless, the major mass of the cloud is represented by those dots, and only a smaller number of low velocity particles keeps its initial size.

Rapid slowing down of fragments in a dense atmosphere brings to a situation when the shock wave overtakes the cloud and reflects from the upper and side walls while the fragments are still in the center of the vessel (Fig. 7.19). The small droplets slow down very rapidly and loose their kinetic energy much faster then the large ones. Thus the large fragments, that had initially much lower velocity, come to overtake the small ones (Fig. 7.19a). those fragments are, actually, the first to collide the upper wall (Fig. 7.20a). The reflected shock wave prevents the small droplets from colliding the upper wall for some time (Fig. 7.20b).

The average wall overpressure profile for the corresponding time ($t = 1.1$ ms) is shown in Fig. 7.21b (curve 10 bar). It is seen, that contrary to the previous case, the overpressure is distributed rather uniformly along the top, bottom and side walls of the containment. Its maximal value could be still found on the top wall at the axis of symmetry. But the maximal average overpressure for the present case is more than an order of magnitude lower.

Figure 7.21a–c illustrates average wall overpressures ($p - p_0$) for one and the same fragments cloud propagating inside the containment, but for different initial gas pressures in the containment. The results show, that on increasing the initial pressure (and density) of the gas filling the containment the maximal overpressures decrease. For the case $p_0 = 1.5$ MPa the overpressures are negligible (Fig. 7.21c) and present only on the bottom and side walls. All the cloud fragments are split into small droplets and slowed down. No disturbance reaches the upper wall.

In case of a fluid-filled containment an overheated expanding gaseous cloud is being formed in the zone of fragments deceleration due to the concentrated energy release. The expansion of the gas-vapour cloud brings to a formation of a diverging shock wave. Reflections of shock waves in fluids from elastic walls take place in the form of the rarefaction waves that lower down the pressure and bring to the formation of the cavitation zones near the walls (Ivashnyov, Ivashneva, Smirnov, 2000). The collapse of those zones usually results in a breakup of the walls. The succession of the processes of internal loading of the fluid-filled containment: energy release in deceleration of fragments, gas-vapour cloud formation and expansion, blast wave propagation, reflection from an elastic shell, cavitation and collapse of cavities — can be described making use of the mathematical models for dynamics of multiphase media accounting for chemical and physical transformations (Smirnov, Zverev, 1992).

7.6. A Concept for Shield Design

The double bumper and multi-shock shield concepts (Christiansen E.L.; Horn J.R. and Crews J.L 1990; Cour-Palais B.G. and Crews J.L. 1990) suggested more than ten years before proved their effectiveness. The above results bring us to a conclusion, that gas-filled containments after some optimization studies could serve a reliable shield protecting the space structures. As multi-sheet shielding concept uses thin shield elements to repeatedly shock the impacting projectile to cause its melting and vaporization, so is the new gas-filled containment shield concept still using continuous effect of pressurized gas to cause fragments slowing down, heating, melting, atomization and evaporation. Besides, using gas-filled bumpers makes it possible to increase the area of the zone of impact energy redistribution including the side and front walls due to the property of gas to transmit pressure in all directions. That is a considerable advantage of the present concept.

The gas-filled bumper shields could be reusable, as the rate of gas phase leakage on depressurization is rather low and the loss of mass is negligible during the characteristic time of impact. The influence of molar mass of the gas phase and other parameters on the rate of impact energy consumption and transformation is to be investigated.

7.7. Conclusions

The worked out mathematical models enable to create a universal approach to orbital breakups modeling. The results for fragments distribution functions and velocities qualitatively coincide with the existing models for different types of breakups and are in a good quantitative agreement with the experimental observations.

Being based on the physical principles, the created breakup models enable to solve inverse problems: to determine the possible breakup scenario using the data of fragments distribution and velocities.

The investigations show, that:

1. number, mass and velocity distributions of fragments depend not only on the total energy of explosion but have a strong nonlinear dependence on the scenario of energy release;
2. there is no unique similarity parameter for the problem, thus simple upscaling of the experiments is impossible.

A detailed modeling of the fragmentation and/or damages in hypervelocity collisions of debris particles with pressurized gas-filled or fluid-filled vessels accounting for the peculiarities of kinetic energy transformation into the energy of blast waves' internal loading brought to developing a new concept of shielding.

The physical and mathematical models for hypervelocity impact on pressurized structures will be very useful in evaluation of potential damages of space vehicles in collisions with debris particles which turn to be the more and more probable.

The effects of energy dissipation in hypervelocity collisions of particles with gas-filled vessels can be used successfully in working out principles for shielding space vehicles by damageable structures transforming the kinetic energy of the impact into the energy of the shield destruction.

MATHEMATICAL MODELS FOR DYNAMICS OF MULTIPHASE MEDIA AND DEFORMABLE STRUCTURES

N.N. Smirnov, A.B. Kiselev and V.F. Nikitin

The methods of multiphase mechanics proved to be very useful for modeling the dynamics of heat release in combustion and detonation of fuel-oxidant mixtures (chemical explosions), hypervelocity interaction of clouds of fragments with pressurized vessels, evolution of polydispersed debris clouds. Those methods were used for solving particular problems discussed in the Chapters 3 and 7 of the present edition. The models described hereinafter were used to obtain the numerical results discussed in the previous chapters. These models could be useful for specialists interested in numerical simulations of dynamical interactions of gaseous, liquid and solid phases with account of physical and chemical transformations.

A.1. Turbulent Flows of Multiphase Polydispersed Chemically Reacting Mixtures

The present section contains the model for polydispersed turbulent flows of thermochemically destructing particles suspended in a gaseous phase. In modeling chemical explosions one could regard the condensed phase as a liquid fuel, or as a solid one. The present model takes into account the possibility for the dispersed phase to contain both liquid and solid components. The model applies both deterministic methods of continuous mechanics of multiphase flows to determine the mean values of parameters of the gaseous phase and stochastic methods to describe the evolution of polydispersed particles in it and fluctuations of parameters. Thus the influence of chaotic pulsations on the rate of energy release and mean values of flow parameters can be estimated. The transport of kinetic energy of turbulent pulsations at the same time obeys the deterministic laws being the macroscopic characteristic.

A.1.1. Mathematical Model for the Gas Phase

The system of equations for gas phase was obtained by Favre averaging of the system of multicomponent multiphase media. We will use the modified *k-epsilon* model to describe the behaviour of the gas phase. The generalization of this model took into account the influence from the other phases as well as the combustion and heat and mass transfer in the gas phase (Smirnov, Nikitin, Legros, 1997).

Averaging by Favre with the $\alpha\rho$ weight (Launder, Spalding, 1972; Pironneau, Mohammadi, 1994) we obtain the following system for the gas phase in a multiphase flow (the averaging bars are removed for simplicity) (Smirnov, Nikitin, Legros, 1997):

$$\partial_t(\alpha\rho) + \nabla\cdot(\alpha\rho\vec{u}) = \dot{M}, \tag{A.1}$$

$$\partial_t(\alpha\rho Y_k) + \nabla\cdot(\alpha\rho\vec{u}Y_k) = -\nabla\cdot\vec{I}_k + \dot{M}_k + \dot{\omega}_k \tag{A.2}$$

$$\partial_t(\alpha\rho\vec{u}) + \nabla\cdot(\alpha\rho\vec{u}\otimes\vec{u}) = \alpha\rho\vec{g} - \alpha\nabla p + \nabla\cdot\tau + \dot{\vec{K}} \tag{A.3}$$

$$\partial_t(\alpha\rho E) + \nabla\cdot(\alpha\rho\vec{u}E) = \alpha\rho\vec{u}\cdot\vec{g} - \nabla\cdot p\vec{u} - \nabla\cdot\vec{I}_q + \nabla\cdot(\tau\cdot\vec{u}) + \dot{E} \tag{A.4}$$

The equations (A.1–A.4) include mass balance in the gas phase, mass balance of *k*-th component, momentum balance and energy balance respectively. We have the following relationships between the terms in the equations (A.1–A.2):

$$\sum_k Y_k = 1, \quad \sum_k \dot{M}_k = \dot{M}, \quad \sum_k \vec{I}_k = 0, \quad \sum_k \dot{\omega}_k = 0.$$

The state equations for gaseous mixture are the following:

$$p = R_g\rho T\sum_k Y_k / W_k, \quad E = \sum_k Y_k(c_{vk}T + h_{0k}) + \frac{\vec{u}^2}{2} + k. \tag{A.5}$$

The *k*-th component mass origination rate $\dot{\omega}_k$ was calculated as a sum of mass production rates ω_{kj} in each *n*-th chemical reaction taking place in a gaseous phase (Smirnov, Zverev, 1992):

$$\omega_k = \sum_{n=1}^{L_\eta} \omega_{kn}.$$

The term responsible for chemical transformations, $\dot{\omega}_k$ is very sensitive to temperature variations, as it is usually the Arrhenius law type function for the reactions' rates. To take into account temperature variations the source term $\dot{\omega}_k$ in the equations (2) was modeled using the Gaussian quadrature technique.

Let us regard the temperature being a stochastic function T with mean $\overline{T}$ and mean squared deviate $\theta = \overline{T'T'}$. Then, the mean value of a function having T as independent variable could be determined as follows:

$$\overline{f(T)} = \int f\left(\overline{T} + \varsigma\sqrt{\theta}\right)P_d(\varsigma)d\varsigma,$$

where ζ is a random value with zero expectation and unit deviate; its probability density function is $P_d(\zeta)$. To estimate the integral, the Gaussian quadrature technique (Hamming, 1962) is applied using the minimal number of terms (namely, three) and assuming $P_d(\zeta)$ to be even. In this case, the formula for $f(T)$ averaging is:

$$\overline{f(T)} = \frac{1}{2\chi^2}f(\overline{T} - \chi\sqrt{\theta}) + \left(1 - \frac{1}{\chi^2}\right)f(\overline{T}) + \frac{1}{2\chi^2}f(\overline{T} - \chi\sqrt{\theta}).$$

The value of χ is of the order of 1; it depends on the particular type of probability distribution function. In case of normal (Gaussian) deviate, it is equal to $\sqrt{3}$ (Gauss-Hermite case).

Therefore, the formula above could be transformed as follows:

$$\overline{f(T)} = \frac{1}{6}f(\overline{T} - \sqrt{3\theta}) + \frac{2}{3}f(\overline{T}) + \frac{1}{6}f(\overline{T} - \sqrt{3\theta}).$$

In our case, the function $f(T)$ is the Arrhenius temperature dependence; the whole average for $\dot{\omega}_k$ is constructed using combinations of these dependencies. Averaged magnitudes for mass fractions, temperature and density were used in the Arrhenius law for $\dot{\omega}_k$ as the dependence of these functions is not as strong as the dependence of temperature.

The turbulent heat flux $\vec{I}_q$ in the equation (A.4) is a sum of two terms:

$$\vec{I}_q = \vec{J}_q + \sum_k (c_{pk}T + h_{0k})\vec{I}_k, \tag{A.6}$$

where $\vec{J}_q$ could be interpreted as turbulent conductive heat flux.

The eddy kinematic viscosity v^t is expressed according to k-epsilon model as $v^t = C_\mu \dfrac{k^2}{\varepsilon}$.

Using the standard k-epsilon model for compressible flows (Pironneau, Mohammadi, 1994), we will model the turbulent fluxes in the following way:

$$\tau = \alpha(\mu + \rho v^t)(\nabla\vec{u} + \nabla\vec{u}^T - (2/3)(\nabla\cdot\vec{u})U) - (2/3)\alpha\rho kU, \tag{A.7}$$

$$\vec{I}_k = -\alpha\rho(D + (v^t / \sigma_d))\nabla \cdot Y_k, \tag{A.8}$$

$$\vec{J}_q = -\alpha\left(\lambda + \sum_k c_{pk}\rho(v^t / \sigma_t)\right)\nabla \cdot T, \tag{A.9}$$

The model is closed then by the equations for k, θ and ε:

$$\partial_t(\alpha\rho k) + \nabla \cdot (\alpha\rho\vec{u}k) = \nabla \cdot (\alpha(\mu + \rho(v^t / \sigma_k))\nabla k) + \tau^t : \nabla\vec{u} - \alpha\rho\varepsilon, \tag{A.10}$$

$$\partial_t(\alpha\rho\varepsilon) + \nabla \cdot (\alpha\rho\vec{u}\varepsilon) = \nabla \cdot (\alpha(\mu + \rho(v^t / \sigma_\varepsilon))\nabla\varepsilon) + (\varepsilon / k)(C_{1\varepsilon}\tau^t : \nabla\vec{u} - C_{2\varepsilon}\alpha\rho\varepsilon) \tag{A.11}$$

$$\partial_t(\alpha\rho\theta) + \nabla \cdot (\alpha\rho\vec{u}\theta) = \nabla \cdot (\alpha(\mu + \sum_k c_{pk}\rho(v^t / \sigma_k))\nabla\theta) + P_\theta + W_\theta - D_\theta, \tag{A.12}$$

where the production terms P_θ, W_θ and the dissipation term D_θ are determined by the following formulae:

$$
\begin{aligned}
P_\theta &= 2\alpha\rho\sum_k c_{pk}\rho(v^t / \sigma_k)(\nabla T)^2, \\
W_\theta &= -\sum_k \overline{\dot{\omega}_k T' h_{0k}}, \\
D_\theta &= C_g\alpha\rho\sum_k c_{pk}\frac{\varepsilon}{k}\frac{\theta}{\theta_m - \theta}.
\end{aligned}
\tag{A.13}
$$

In deriving the production W_θ due to chemistry the Arrhenius law for chemical transformations was assumed. To calculate the averaged term $\dot{\omega}_k$ the Gaussian quadrature technique was applied similar to the technique used to calculate averaged Arrhenius terms. It gives the following result for a single Arrhenius temperature dependence term:

$$\overline{T'A(T)} = \theta\frac{A(\bar{T} + \sqrt{3\theta}) - A(\bar{T} - \sqrt{3\theta})}{2\sqrt{3\theta}}.$$

The dissipation term D_θ is constructed with the following assumption: the squared temperature deviate cannot exceed its maximal possible value θ_m, because the value of $T = \bar{T} + T'$ cannot be negative. However, production terms do not grant the presence of such a boundary. To grant it with sure, we incorporate the multiplier $1/(\theta_m - \theta)$ into the dissipation term (the other multipliers are standard, see [Philip, 1991]). In order to estimate the value of θ_m, we should take into account that the probability for the deviate value to exceede 2 times the mean deviate value is less than 1% for the normal distribution. Also, we should take into account that the mean temperature deviate in experiments published by Philip (1991) did not exceed half of the maximal mean temperature. With this, we estimate θ_m as follows:

$$\theta_m = \bar{T}^2 / 4.$$

The constants in (A.10–A.12) take the following standard values (Launder, Spalding, 1972; Pironneau, Mohammadi, 1994):

$$C_\mu = 0.09, \quad C_{1\varepsilon} = 1.45, \quad C_{3\varepsilon} = 1.92, \quad \sigma_d = 1, \quad \sigma_t = 0.9, \quad \sigma_k = 1, \quad \sigma_\varepsilon = 1.13.$$

The dissipation constant C_g in (13) could be determined based on the experiments (Philip, 1991):

$$C_g = 3.8.$$

One could see that the equations (A.10–A.12) do not contain terms responsible for particulate phase contribution to turbulence energy growth. This is due to the direct stochastic modeling of the particulate phase: the influence of the last on the gas phase leads to stochastic behavior of the momentum source terms $\bar{K}$. These terms affect the averaged gas phase velocity in the stochastic manner and therefore the source term $\tau' : \nabla \bar{u}$ is also affected. Consequently, the influence of the stochastic behavior of the particulate phase is taken into account by that term in the right-hand side of both equations (A.10–A.12). Thus with this approach additional terms should not be included into the equations (A.10–A.12) not to take twice the impact of the particulate phase into account.

With the equations (A.10–A.12) and expressions (A.7–A.9) the model to be closed we need expressions for mass, momentum and energy fluxes from the other phases described below and the initial and boundary conditions.

A.1.2. Dispersed Phase Modeling

The motion of polydispersed particulate phase is modeled making use of a stochastic approach. A group of representative model particles is distinguished. Motion of these particles is simulated directly taking into account the influence of the mean stream of gas and pulsations of parameters in gas phase. Properties of the gas flow: the mean kinetic energy of turbulence and the rate of pulsations decay — make it possible to simulate the stochastic motion of particles determining the frequency of changes of pulsations under the assumptions of the Poisson flow of events.

We model a great amount of real particles (for instance, liquid droplets) by an ensemble of model particles (their number is of the order of thousands). Each model particle is characterized by a vector of values, representing its location, velocity, mass and other properties. Here we regard the following vector, determined for each model particle:

$$\{N, m, m_s, \vec{r}, \vec{v}, \vec{w}, \omega, T_c\}_i, \quad i = 1, ..., N_p. \tag{A.14}$$

The values of N_i are initialized to satisfy the equality $\sum_i N_i m_i = M_p$, and are not changed in the process of calculations. When a particle is burnt out, its mass m_i is set to zero, and this particle is excluded from calculations.

In case of fragmentation of particles or atomization of droplets the number N_i increases and the mass decreases so as to preserve the equality with the total mass represented by a model particle by that time.

Modeling the particles phase therefore is split into two stages: first stage is the evaluation of the vector (A.14) for each model particle, and the second stage is the evaluation of the particulate phase volumetric share $\alpha_2 = 1 - \alpha$ and fluxes $\dot{M}_k, \dot{M}, \vec{K}, \dot{E}$ used in the equations (A.1–A.4) describing the gas phase behavior.

Hereinafter in this section we will omit the subscript "i" denoting the belonging of parameters to the i-th model particle for simplicity. To model the motion of particles we use the approach described in details in (Smirnov, Nikitin, Legros, 1997). The laws for particle's motion are:

$$m\frac{d\vec{v}}{dt} = m\vec{g} - \frac{m}{\rho_c}\nabla p + \vec{f}_d, \quad \frac{d\vec{r}}{dt} = \vec{v}, \tag{A.15}$$

where the force affecting the particle consists of gravity and Archimedus forces, resistance force and a stochastic force resulting from interaction with turbulized gas; the gas phase parameters are supposed to be evaluated in the vicinity of the model particle.

Traditionally the stochastic force is modeled according to Langevin: white noise is applied as a derivative of a Wiener process. This approach for example is used to describe the Brownian motion. But in our case we regard the particles much larger than those affected significantly by the Brownian motion, and the chaotic change of their trajectories has a different nature. The Brownian particles are affected by the molecular pressure fluctuations, and the particles much larger than the Brownian ones — by the influence of turbulent pulsations of gas stream in the vicinity of each particle. The effect of gas velocity pulsations is transferred to the particulate phase mainly via the resistance force, which depends on the difference of gas and particle velocities. So it can be useful in our case to determine the stochastic component of the force acting on each model particle by means of introducing the stochastic component of the gas velocity in the neighborhood of the particle. In this case the force $\vec{f}_d$ will contain both the average and the stochastic components.

Using the notation $\vec{w}$ for the stochastic component of gas velocity in the vicinity of the model particle, we consider the following law for the force $\vec{f}_d$:

$$\vec{f}_d = \frac{1}{2}C_d(\mathbf{Re})\rho S(\vec{u} + \vec{w} - \vec{v})|\vec{u} + \vec{w} - \vec{v}|,$$

the coefficient C_d being determined by the following expression (Smirnov, Zverev, 1992):

$$C_d = 24 \cdot \text{Re}^{-1}(1 - 0.183\sqrt{\text{Re}} + 1.72 \cdot 10^{-2}\,\text{Re})\beta_s^{\frac{4}{5}}\kappa,$$

$$\beta_s = \sqrt{\frac{\rho}{\rho_s}\left(2 - \frac{\rho}{\rho_s}\right)}; \quad \kappa = \left(\frac{\rho_w}{\rho}\right)^{\frac{1}{5}}\left(\frac{\mu_w}{\mu}\right)^{\frac{1}{5}}; \quad \rho_w = \rho(T_w, p); \quad \mu_w = \mu(T_w, p)$$

$$\frac{\rho_s}{\rho} = \begin{cases} \left(1 + \dfrac{\gamma - 1}{2}M\right)^{\frac{1}{\gamma-1}}; & M < 1 \\[3mm] \dfrac{(\gamma + 1)M^2}{(\gamma - 1)M^2 + 2}\left[1 + \dfrac{\gamma - 1}{2} \cdot \dfrac{(\gamma - 1)M^2 + 2}{2\gamma M^2 - (\gamma - 1)}\right]^{\frac{1}{\gamma-1}}; & M > 1 \end{cases}$$

where the local Reynolds and Mach numbers are:

$$\mathbf{Re} = \frac{d|\vec{u} + \vec{w} - \vec{v}|}{\nu}, \quad M = \frac{|\vec{u} + \vec{w} - \vec{v}|}{a}, \quad |\vec{u} + \vec{w} - \vec{v}| = v_{rel}.$$

To model the vector $\vec{w}$, the following technique is used. Let the particle enter the new pulsation at the time moment t^*. The new pulsation velocity vector $\vec{w}_t$ is described by the expression:

$$\vec{w} = \sqrt{2k} \cdot \vec{\xi},$$

where k is the kinematic energy of pulsations, its value being taken as the gas phase turbulent energy described by *k-epsilon* model (A.10–A.12) and evaluated in the vicinity of the particle. The random vector $\vec{\xi}$ is supposed to be distributed normally with the unit dispersion, so that its density of distribution is equal to:

$$P_d(\vec{\xi}) = \frac{1}{(2\pi)^{3/2}}\exp\left(-0.5 \cdot \left|\vec{\xi}\right|^2\right).$$

After the pulsation component of the gas stream velocity in the vicinity of the model particle is determined, it is kept constant for this particle until the event of the pulsation change occurs. This stochastic stream of events is supposed to have no pre-history. Thus it is modeled by the Poisson law. According to this law, the probability of the pulsation-change event within the time interval Δt is the following:

$$P_{\Delta t} = 1 - \exp(-\Delta t/t^c),$$

where t^c is the characteristic time of the Poisson flow of events. In our model we consider the value of t^c being dependent on the turbulent characteristics of the gas flow in the vicinity of the model particle. We use the following rule:

$$t^c = \begin{cases} k/\varepsilon & |\vec{w}| < 3\sqrt{2k} \\ 0 & |\vec{w}| \geq 3\sqrt{2k} \end{cases}.$$

The last expression has the following physical background: if the velocity of pulsation in the vicinity of the particle is less than nearly maximal possible for the normal law of its distribution then the characteristic time of its change is taken equal to the characteristic time of turbulence decay. If it is abnormally large then it must be changed immediately so the value of f is set to be zero and therefore the probability of the gas stream pulsation change $P_{\Delta t}$ is set to be 1. That happens when particles enter less turbulized zones of the flow.

The above concept for stochastic modeling of the particulate phase behavior rigorously based on the gaseous phase local characteristics developed by deterministic methods was first suggested in the paper (Dushin, Nikitin, Smirnov, 1993) and since then is very widely used by many authors for turbulent combustion simulations.

The decay of mass from a droplet/particle is evaluated by the following expression:

$$\frac{dm}{dt} = -\dot{m},$$ (A.16)

while the decay of volume is determined by the decay of mass of solid component only. Thus the extraction of volatiles causes sometimes the decay of mean density of particles. The expression for $\dot{m}$ will be given below.

The internal energy of the particle changes due to heat exchange with the surrounding gas and due to possible phase transitions or chemical reactions on the particle's surface:

$$m\frac{de}{dt} = q + Q_s,$$ (A.17)

where q is a heat flux between the gas and the particle:

$$q = \begin{cases} \pi d \lambda \mathbf{Nu}(T - T_s), & \mathrm{Re} < 10^3; \\ \pi d^2 \rho v_{rel} \mathbf{St}(H_r - H_w), & \mathrm{Re} > 10^3; \end{cases}$$

$$H_r = c_p T + r_a \frac{v_{rel}^2}{2}; \quad H_w = c_{pw} T_w.$$ (A.18)

The Nusselt, Stanton numbers and the accomodation coefficient for the i-th particle (droplet) are determined by the formulas:

$$\mathbf{Nu} = 2 + 0.16 \cdot \mathbf{Re}^{2/3}\mathbf{Pr}^{1/3}, \quad \mathbf{St} = \frac{1}{2}C_d\mathbf{Pr}^{-2/3}, \quad r_a = \mathbf{Pr}^{1/3},$$

where the Reynolds number $\mathbf{Re}$ is determined accounting for the influence of turbulent pulsations of velocity, which induce oscillations in the heat flux between gas and particles.

The internal energy for multi-component particles is determined by the formula:

$$e = \sum_{j=1}^{L} (c_j T_s + h_j^0)Y_j,$$

where c_j — heat capacity of the j-th component within the particle and h_j^0 — the enthalpy of the gas phase chemical reactions. The heat release or absorption on the surface of the particle due to chemistry or phase transitions can be determined by the formula:

$$Q_s = \sum_{j=1}^{L} \dot{m}_j h_j^s,$$

where $\dot{m}_j$ is the mass rate of consumption (or extraction) of the j-th component from the particle; h_j^s is the enthalpy of surface chemical reactions or phase transformations.

Accounting for droplets atomization

In case the dispersed phase contains liquid droplets the dynamic interaction with the gaseous flow could bring to an instability of the interface and atomization of droplets. The criterion for liquid droplets instability is that of the critical Weber number (Azzopardi, Hewitt, 1997):

$$We = \frac{\rho v_{rel}^2 d}{\sigma},$$

where σ is the surface tension at the interface. On exceeding the critical value of the Weber number droplets breakup due to vibrational instability takes place. On essentially surpassing the critical Weber number other mechanisms start playing essential roles in the breakup process that brings to formation of fine mist. These main characteristics of the atomization process could be taken into account by the following approximate formula (Smirnov, Zverev, 1992; Smirnov, Tyurnikov, 1994) determining mean diameters of droplets d_a originating in breakups of initial droplets (diameter d):

$$d_a = \begin{cases} d = \left(\dfrac{6\alpha_2}{\pi n}\right)^{\frac{1}{3}}, & We < We_*; \\[3mm] \dfrac{d We_*}{We}, & We_* \leq We \leq We_{**}; \\[3mm] d_*, & We > We_{**}; \end{cases} \tag{A.19}$$

where n is the number of droplets per volume unit, the critical Weber numbers are determined as follows (Smirnov, Zverev, 1992; Azzopardi, Hewitt, 1997):

$$\begin{aligned} We_* &= 12(1 + Lp^{-0.8}) \\ We_{**} &= 350 \end{aligned},$$

where the Laplace number $Lp = \dfrac{d\rho_c \sigma}{\mu_c^2}$.

To determine the mean diameter of droplets d_* after the breakup of a type of an explosion ($We > We_{**}$) one needs to evaluate the part of the accumulated by a droplet energy spent for the breakup E_* (a more detailed description is present in the chapter 7). The assumption, that the breakup energy was spent for the formation of new free surface makes it possible to evaluate the number N and the mean diameter d_* of the formed droplets:

$$N = \left(1 + \frac{E_*}{\sigma \pi d^2}\right)^3; \quad d_* = \frac{d}{1 + \frac{E_*}{\sigma \pi d^2}}.$$

The breakup energy could be evaluated as the difference between the work of the drag forces separating small droplets from the initial one, and the kinetic energy of fragments' scattering:

$$E_* = A_{drag} - \sum_{i=1}^{N_*} \frac{m_i v_{i*}^2}{2}.$$

Assuming that the initial droplet is split into N_* equal droplets $\left(N_* = \frac{d^3}{d_*^3}\right)$ having equal velocities of radial expansion of the cloud v_* and the separation of droplets takes place after the droplet is moved away at a distance $\sim d_*$, one obtains the following formulas:

$$A_{drag} = \frac{1}{8} N_* \rho C_d v_{rel}^2 \pi d_*^2 d_*;$$

$$d_* = \frac{d}{1 + \frac{1}{4}\left(\frac{1}{2} C_d \rho v_{rel}^2 - \frac{1}{3} \rho_c v_*^2\right)\frac{d}{\sigma}}.$$

The mean velocity of the cloud expansion v_* could be evaluated based on the condition of matching the two formulas (A.19) for d_a at $We = We_{**}$. The reason to perform that matching is that both formulas for breakup regimes were obtained from experiments, thus, indirectly the expansion of the cloud of droplets after the breakup should have been taken into account. On the other hand, the dependence of characteristic droplets diameters on the Weber number should be continuous. Then, finally one obtains

$$d_* = \frac{d We_*}{\frac{1}{8} C_d (We - We_{**}) We_* + We_{**}}.$$

In modeling droplets breakup in a gas flow the inertia of the process should be taken into account. Fragmentation does not take place instantaneously: it needs time for small droplets to separate from the initial one, i. e. it needs a definite time for the liquid bridges between the droplets to be established, elongated and broken. The first order estimate of the breakup time gives the following formula:

$$t_* = \frac{d}{v_{rel}} \frac{\mathbf{We}_*}{\mathbf{We}} \left(1 + \frac{3}{8} C_d \frac{\rho}{\rho_c} \left(1 - \frac{4}{C_d \mathbf{We}_*}\right)\right).$$

Non-equilibrium effects in phase transformations

The mass exchange processes between a particle and the gaseous phase can take place due to phase transitions, devolatilization of dust particles, chemical transformations at the interface, etc. For evaporating components the rate of mass extraction is

$$\dot{m}_j^{(1)} = \frac{q_j}{h_j^3}, \tag{A.20}$$

where q_j is the heat flux spent for evaporation, h_j^s is the specific enthalpy of phase transformations. The volatiles extraction could be described by the following kinetics:

$$\dot{m}_j^{(2)} = (m - m_3) A_j^{(2)}(T_s) \tag{A.21}$$

The rate of heterogeneous chemical transformations on the surface of the particles is strongly affected by both chemical kinetics and diffusion of gaseous components (reagents/ reaction products) to/from the interface (Smirnov, Zverev, 1992):

$$\dot{m}_j^{(3)} = K_j \log\left(1 + Y_k \frac{v_j W_j}{v_k W_k}\right)^{y_j}; \tag{A.22}$$

where

$$K_j = \pi d^2 \left(\frac{v_k W_k}{v_j W_j A_j^{(3)}(T_s)} + \frac{d}{\rho D \mathbf{Nu}_D}\right)^{-1};$$

$$y_j = \dot{m}_j / \dot{m};$$

$A_j^{(2,3)}(T_s)$ — the Arrhenius functions for heterogeneous kinetics depending on the surface temperature

$$A_j(T_s) = \begin{cases} G_j T^{g_j} \exp\left(-\frac{E_{A_j}}{R_g T}\right); & T > T_{*_j} \\ 0; & T < T_{*_j} \end{cases}$$

The subscript k in the formulas (A.22) denotes the number of the gaseous components participating in the reaction with the j-th condensed component. The diffusion Nusselt number is determined by the formula:

$$Nu_D = 2 + 0.16 \cdot Re^{2/3} Sc^{1/3}, \quad Sc = \frac{\mu}{\rho D}.$$

The equations (A.22) are the generalization of the solutions obtained in (Smirnov, Zverev, 1992) for quasisteady heterogeneous diffusive combustion of spherical particles. Under the condition $Y_k W_j v_j \ll y_j W_k v_k$ the respective formula (A.22) could be essentially simplified:

$$\dot{m}_j = K_j \log\left(1 + \frac{Y_k W_j v_j}{W_k v_k}\right).$$

The mass flux due to evaporation is not as easy to be determined as one could think on regarding the formula (A.20). The heat flux q_j spent for evaporation is, actually, the function of the state of the interface. Non-equilibrium effects could strongly affect the process under certain conditions.

Mathematical models for individual droplets combustion incorporated in polydispersed mixtures modeling, are usually based on the assumptions of the equilibrium character of phase transitions. Comparison of theoretical and experimental data shows, that this assumption being undoubtedly valid for large droplets and flat surfaces, brings to essential errors for small droplets. As on burning out a droplet passes all the size-stages, it is necessary to take into account the non-equilibrium effects in phase transitions. A detailed analysis of the problem could be found in (Smirnov, Kulchitsky, 2000). Here we'll give a brief coverage of the results. A new dimensionless parameter characterizing the deviation of phase transition from the equilibrium was introduced: $I_e = \dfrac{D}{r_0 \delta}\sqrt{\dfrac{2\pi}{R_g T_a}}$, where r_0 — the droplet radius, D — diffusion coefficient, δ — accommodation coefficient, T_a is the characteristic temperature of the surrounding gas.

The rate of droplets evaporation is characterized by a dimensionless Peclet number: $Pe = \dfrac{\dot{m} r_0}{\rho D}$, where $\dot{m}$ is the mass rate of evaporation, ρ is the average density of gaseous phase.

The results of comparison of non-equilibrium and quasi-equilibrium Peclet numbers for droplet evaporation show that the rate of evaporation strongly depends on the value of the I_e parameter. On increasing I_e the deviations of the evaporation rates from those predicted by the equilibrium model turn to be larger and larger. Thus in order to have adequate estimates for small droplets evaporation rate one needs to use the non-equilibrium model for $I_e > 1$. The non-equilibrium solution of the problem is also free of the major drawback of the quasi-equilibrium one. The quasi-equilibrium solution loses validity on decreasing the droplet radius, because the evaporation rate and vapors velocities tend to infinity. The non-equilibrium solution, at the same time, gives final values for flow velocities, that turn to be subsonic as long as the condition $W_1 < 2\pi W_W \gamma_W$ is satisfied, where W_1; W_W — molar masses of the evaporating component and the ambient mixture; γ_W — the polytropic constant for the mixture.

Thus the obtained non-equilibrium solution does not lose physical sense in all the range of governing parameters variation. It shows that on decreasing droplets radii the deviations of the system from the equilibrium state could turn to be enormous that results in essential difference in the rates of evaporation. A universal criterion is introduced characterizing the deviation of a system from an equilibrium state.

A.1.3. Fluxes from Model Particles and Their Recalculation into Gas Phase Equations Source Terms

The rate of particle's mass decrease, is calculated by the formula (subscript i is omitted for simplicity):

$$\dot{m} = \sum_{j=1}^{L} \dot{m}_j.$$

Momentum and energy fluxes are calculated as follows:

$$\vec{k} = \dot{m}\vec{v} - \vec{f}_r, \quad \dot{e} = q_s + \dot{h} + 0.5\dot{m}|\vec{v}|^2 - (\vec{v}\cdot\vec{f}_d), \quad \dot{h} = \sum_{j=1}^{L} \dot{m}_j(e_j + h_j^0);$$

q_s — heat flux coming inside the particle.

Fluxes $\dot{M}_k$, $\dot{M}$, $\vec{K}$, to gas phase as well as the volumetric share of particles phase are obtained evaluating correspondent fluxes from model particles and volume of model particles. The procedure of recalculating (Smirnov, Nikitin, Legros, 2000) is the following.

For each grid node n with the volume Ω_n attached to it, we evaluate at the first stage:

$$F_n^1 = \frac{1}{|\Omega_n|} \sum_{i:(x_i,r)\in\Omega_n} f_i \ ,$$

where F is an element of the set defined on the grid: $F \in \{\alpha_2, \dot{M}, \dot{M}_k, \dot{K}_x, \dot{K}_r, \dot{E}\}$, and f is an element of the corresponding set defined on model particles via the values from the particle's vector (A.14) and fluxes evaluated from it: $f \in \left\{\omega, \dot{m}, \dot{m}_k, \dot{k}_x, \dfrac{y\dot{k}_y + z\dot{k}_z}{\sqrt{y^2 + z^2}}, \dot{e}\right\}$.

Then we apply the procedure smoothing the fields of F^1 in order to ensure stability of calculations. The detailed description of the algorithm is given in (Smirnov, Nikitin, Legros, 1997).

The method is based on the successive application of the following procedure to the adjacent nodes n and m:

$$F_n^2 = F_n^1 + \kappa(F_m^1 - F_n^1)\sqrt{|Q_m|/|Q_n|}$$
$$F_m^2 = F_m^1 + \kappa(F_n^1 - F_m^1)\sqrt{|Q_n|/|Q_m|}$$

(A.23)

The procedure described above (A.23) conserves the volumetric integral of F because

$$F_n^2|Q_n| + F_m^2|Q_m| = F_n^1|Q_n| + F_m^1|Q_m|.$$

If the value of κ is kept in the interval [0, 0.5], then the procedure (A.23) applied to each pair of adjacent nodes of the grid will lead to stable smoothing of the primary distribution. This procedure can be applied successively and the way of nodes encounting may vary from one node to another. Also, it can be repeated as many times as it is needed to reduce noise in the distribution. In our calculations we used 4 global iterations of this procedure for each flux and particles volume share, and the κ multiplier was set to be 0.1 to avoid extra numerical diffusion.

A.1.4. Numerical Modeling Techniques

Each time step contains the calculations of model particles motion, determining fluxes from particles to the gas phase and recalculating them to the grid. Motion of each particle is calculated using an iterative implicit algorithm independently for each particle. Then two iterations to calculate gas dynamics parameters are undertaken accounting for fluxes from the particulate phase. At each iteration the space splitting in x and r coordinates is used as well as the splitting in three physical processes: chemistry, source terms and turbulent energy production (local part of the equations), convection (hyperbolic part), diffusion (parabolic part) (Smirnov, Nikitin, 1997).

The operator splitting techniques represents the general operator $L(\Delta t)$ transferring the parameters vector $\vec{P}^n$ to the next time step $\vec{P}^{n+1}$:

$$\vec{P}^{n+1} = L(\Delta t)\vec{P}^n$$

in the following form:

$$L(\Delta t) = L_x(\Delta t_x)L_r(\Delta t_r)L_r(\Delta t_r)L_x(\Delta t_x) ,$$

or

$$L(\Delta t) = L_r(\Delta t_r)L_x(\Delta t_x)L_x(\Delta t_x)L_r(\Delta t_r) .$$

(A.24)

The sequence of operators in (A.24) yields the condition of symmetry. To yield the condition of time steps ensuring the second order of approximation along with (A.24), we have:

$$\Delta t = 2\Delta t_x = 2\Delta t_r.$$

Both sequences in (A.24) can represent the general operator $L(\Delta t)$. It could be effective to alter sequences at each time step. Each operator itself, L_x and L_r is also split into three parts: the parabolic part L_x^P, L_x^P; the hyperbolic part L_x^H, L_x^H and the local part L_x^s, L_x^s. The local part is solved implicitly using an iterative algorithm independently for each grid node. The hyperbolic part was solved using explicit FCT techniques (Oran, Boris, 1987); the parabolic part was solved implicitly using 3-diagonal matrix solvers for linear equations (Smirnov, Nikitin, 1997).

The time step is calculated using the CFL criterion:

$$(\Delta t_x)^{-1} = \max\left(\frac{|u_x|+a}{\Delta x}\right), \quad (\Delta t_r)^{-1} = \max\left(\frac{|u_r|+a}{\Delta r}\right), \quad \frac{\Delta t}{2} = C \cdot \min(\Delta t_x, \Delta t_r),$$

where a is the sound velocity in gas. The Courant number C is to be assumed $C = 0.2$, based on our experience.

To ensure the necessary precision of calculating fluxes from the particulate phase the following condition should be satisfied:

$$\frac{N_p}{N_x N_r} > C_N. \qquad (A.25)$$

The physical meaning of the criterion C_N in (A.25) is the following: C_N is the minimal average number of model particles per a grid node guaranteeing sufficient accuracy of fluxes between phases evaluation. The decrease of the number N_i of real particles represented by a model one increases the N_p value thus increasing the accuracy of fluxes evaluation.

The criterion

$$\frac{N_i}{\sum_i N_i} < C_\Sigma \qquad (A.26)$$

should be satisfied as well. Numerical verification of the code showed that the criterion (A.26) being set $C_\Sigma = 2 \cdot 10^{-4}$ and the criterion (A.25) being satisfied variations of N_i did not bring to discrepancies exceeding 5%.

The gas dynamic part of the scheme was validated comparing with standard solutions (Smirnov, Nikitin, 1997). The physical model for gas-particles flows and the phenomenological laws for phases interactions were validated by comparing the results of numerical modeling of multi-phase hydrocarbon-air mixtures combustion with the results of shock-tube experiments (Smirnov, Zverev, 1992; Smirnov, 1988; Smirnov, Tyurnikov, 1994).

A.1.5. Nomenclature

Latin symbols

A_r – Arrhenius term for r-th reaction depending on temperature T,

a [m/s] — sound velocity in gas,

C – the Courant number,

C_d — the drag coefficient,

C_N, C_Σ – particulate phase computational factors,

c_μ, $c_{1\varepsilon}$, $c_{2\varepsilon}$ — k-epsilon model constants,

c_{pk}, c_{vk} [J/kg·K] — specific heat capacity of k-th gas component at constant pressure and volume respectively,

D [m^2/s] — overall laminar diffusive coefficient,

d_i [m] — diameter of i-th model particle,

E [J/kg] — specific energy of fluid,

E_A [J] — activation energy,

$\dot{E}$ [J/(m^3·s)] — specific energy flux to gas phase,

$\dot{e}_i$ [J/s] — energy flux from i-th model particle,

F_n – notation for any magnitude

recalculated from the model particles ensemble to the grid node n,

f_i – notation for any magnitude to be recalculated from i-th model particle to the grid,

f_{di}, f_{ri} — [N] total and resistance force acting on i-th model particle,

$\vec{g}$ [m/s^2] — gravity acceleration vector,

$\vec{I}_k$ [kg/(m^{2}s)] — turbulent diffusive flux to k-th component,

$\vec{I}_q$ [J/(m^{2}s)] — turbulent energy flux,

$\vec{J}_q$ [J/(m^{2}s)] — turbulent conductive heat flux,

h_{0k} [J/kg] — specific internal energy of k-th gas component derived to $T = 0$,

$\dot{\vec{K}}$ [kg/(m^{2}s^2)] — specific momentum flux to gas phase vector,

k [J/kg] — turbulent kinematic energy,

$\dot{\vec{k}}_i$ [kg·m/s^2] — momentum flux from i-th model particle,

L – number of components within the particulate phase,

$\dot{M}$ [kg/(m^{3}s)] — specific mass flux to gas phase,

$\dot{M}_k$ [kg/(m^{3}s)] — specific mass flux to k-th component from the dust phase,

M_p [kg] — total initial mass of the particulate phase,

m_i [kg] — mass of i-th model particle,

m_{si} [kg] — mass of solid in i-th model particle,

N_x, N_r – number of grid nodes along the axial and radial axes respectively,

N_i – number of real particles represented by the i-th model particle,

N_p – total number of model particles,

$\mathbf{Nu}_i$ – Nusselt number for the i-th model particle,

$\bar{n}$ – normal vector to a wall,

Q_{si} [J/s] — heat release on the surface of the i-th model particle,

q_i [J/s] — heat flux to the i-th model particle,

$P_d(\xi)$ — probability density of a random value ξ,

$P_{\Delta l}$ – probability of particle's changing to a new pulsation,

$\mathbf{Pr}$ — Prandtl number,

p [Pa] — pressure,

R [m] — radius of the cylindrical vessel,

$\mathbf{Re}_i$ — Reynolds number for i-th model particle,

$R_g = 8.31$ J/(mol·K) — universal gas constant,

$\bar{r}_i$ [m] — radius-vector of i-th particle in Descartes coordinate system,

S_i [m^2] — the middle cross section area of the i-th particle,

St — Stanton number,

T [K] — temperature of gas,

T_{ic} [K] — temperature of i-th model particle,

T_w [K] — temperature of gas near the surface of a particle,

t [s] — time,

t_0 [s] — time of ignition,

U – unit tensor of the 2nd range,

$\bar{u}$ [m/s] — fluid velocity vector,

$\bar{v}_i$ – velocity of i-th model particle,

W_k [kg/mol] molar mass of k-th gas component,

$\bar{w}_i$ [m/s] — random term in the velocity of fluid in the neighbourhood of i-th model particle,

X [m] — length of the cylindrical vessel,

Y_k – mass fraction of k-th gas component,

Y_{ij} – mass fraction of j-th component in the i-th model particle.

Greek symbols

α – volumetric fraction of the gas phase,

α_2 – volumetric fraction of the particulate phase,

$\gamma = C_p/C_v$ – polytropic constant,

ΔH – specific chemical energy release per fuel,

ε [W/kg] — turbulent energy decay,

λ [W/(m^{2}K)] — effective laminar thermoconductivity,

μ [Pa·s] — effective laminar viscosity,

ν [m^2/s] — molecular kinematic viscosity,

ν^t [m^2/s] — turbulent kinematic viscosity,

$\vec{\xi}$ – random unit vector, described in 2.2,

$\mu = 3.14159...$,

ρ [kg/m^3] — gas density,

ρ_c [kg/m^3] –density of the condensed phase,

σ [kg/s^2] — surface tension,

σ_d, σ_t, σ_ε – k-epsilon model constants,

τ [kg/(ms^2)] — turbulent viscosity tensor,

τ^t [kg/(ms^2)] — Reynolds tensor term within effective viscosity tensor τ,

Ω_n [m^3] — volume attached to n-th node of the grid,

ω_i [m^3] — volume of i-th model particle,

$\dot{\omega}_k$ [kg/(m^{3}s)] — specific mass flux to k-th component via chemical reactions in gas phase.

Other notations

∂_t – partial derivative with respect to time,

$\nabla \cdot \vec{a}$– divergence of the vector $\vec{a}$,

$\nabla \cdot \tau$– divergence of the tensor τ,

∇p– gradient of the scalar p,

$\nabla \vec{a}$– tensor with components $\nabla_i a_j$,

$\vec{a} \otimes \vec{b}$– tensor with components $a_i b_j$,

τ^T – transposed tensor τ,

$\tau{:}\theta$ – double product of the 2nd range tensors.

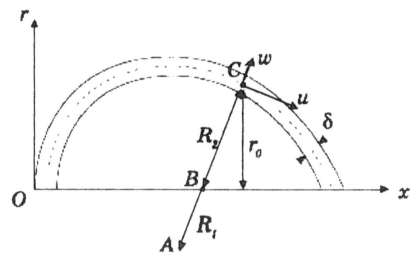

Fig. A1.

A.2. Mathematical Model for Dynamical Deforming and Breakup of Thin-walled Elastoviscoplastic Shell

Regard a typical axisymmetrical thin-walled shell simulating a closed reservoir or fuel tank (Fig. A.1). Two systems of coordinates are used to describe the shell motion. They are: the laboratory co-ordinate system (x, r) with the x-axis along the axis of symmetry of the shell and r-axis in the radial direction; and the moving (accompanying) system of co-ordinates connected with the shell (Kiselev, 1981; Kiselev, Maksimov, 1982, 1984, 1995). The last is introduced so that its s-axis coincides with a positive tangent, z-axis — with a positive normal to the middle surface of the shell. The laboratory system is used to describe the current geometry of the shell and to introduce a moving local basis. The governing system of equations modeling the shell behavior is formulated in the moving accompanying system of co-ordinates.

The governing system of equations for a shell within the frames of Timoshenko approximations has the following form (Kiselev, Maksimov, 1995):

$$\rho \frac{\partial^2 u}{\partial t^2} = \frac{1}{\delta r_0} \frac{\partial(\delta r_0 T_1)}{\partial s} - \frac{1}{r_0} \frac{\delta r_0}{\partial s} T_2 + \frac{Q}{R_1} + \frac{1}{\delta}\left[P_s\left(\frac{\delta}{2}\right) + P_s\left(-\frac{\delta}{2}\right)\right],$$

$$\rho \frac{\partial^2 w}{\partial t^2} = \frac{1}{\delta r_0} \frac{\partial(\delta r_0 Q)}{\partial s} - \left(\frac{T_1}{R_1} + \frac{T_2}{R_2}\right) - \frac{1}{\delta}\left[P_n\left(\frac{\delta}{2}\right) - P_n\left(-\frac{\delta}{2}\right)\right],$$

$$-\rho \frac{\delta^2}{12} \frac{\partial^2 \varphi}{\partial t^2} = \frac{1}{\delta r_0} \frac{\partial(\delta r_0 M_1)}{\partial s} - \frac{1}{r_0} \frac{\partial r_0}{\partial s} M_2 - Q + \frac{1}{2}\left[P_s\left(\frac{\delta}{2}\right) - P_s\left(-\frac{\delta}{2}\right)\right],$$

(A.27)

where M_j, T_j, Q $(j = 1,2)$ — bending moments, normal forces and transverse (crossing) forces; R_1, R_2 — main curvature radii of the middle surface of the shell; r_0 — the distance between the middle surface of the shell and its axis; ρ — density of the material; $P_s\left(\pm\delta/2\right)$, $P_n\left(\pm\delta/2\right)$ — tangential and normal loading forces on the bounding surfaces of the shell $z = \pm\delta/2$. Positive directions for P_s are that coinciding with the s-axis, normal loading P_n forces are considered positive if they cause compression of the shell (internal pressure, for example); u, w — displacements along tangent and external normal to the middle surface; φ — rotation angle for the normal crossection; s — Lagrangian coordinate along the meridian; $\delta = \delta(s)$ — the shell thickness measured along the normal to the middle surface.

All the geometrical parameters in (A.27), but for $\delta = \delta(s)$ are functions of time t. The shell thickness is assumed to be small: $\dfrac{\delta}{R_1} \ll 1$; $\dfrac{\delta}{R_2} \ll 1$. Forces and moments per area unit are determined by the following integrals:

$$T_j = \frac{1}{\delta}\int\limits_{-\delta/2}^{\delta/2} \sigma_{jj}dz, \quad Q = \frac{1}{\delta}\int\limits_{-\delta/2}^{\delta/2} \sigma_{1z}dz, \quad M_j = \int\limits_{-\delta/2}^{\delta/2} \sigma_{jj}zdz. \tag{A.28}$$

Rates of meridian tangential deformation and transverse shear deformation could be expressed via rates of displacements u, w and normal crossection rotation angle φ as follows:

$$\dot{\varepsilon}_1 = \dot{\varepsilon}_1^0 - z\dot{k}_1, \quad \dot{\varepsilon}_2 = \dot{\varepsilon}_2^0 - z\dot{k}_2, \quad \dot{\varepsilon}_{1z} = \dot{\varepsilon}_{1z}^0;$$

$$\dot{\varepsilon}_1^0 = \frac{\partial \dot{u}}{\partial s} + \frac{\dot{w}}{R_1}, \quad \dot{\varepsilon}_2^0 = \frac{\dot{u}}{r_0}\frac{\partial r_0}{\partial s} + \frac{\dot{w}}{R_2}, \quad \dot{\varepsilon}_{1z}^0 = \frac{1}{2}\left(\frac{\partial \dot{w}}{\partial s} - \frac{\dot{u}}{R_1} - \dot{\varphi}\right);$$

$$\dot{k}_1 = \frac{\partial \dot{\varphi}}{\partial s}, \quad \dot{k}_2 = \frac{\dot{\varphi}}{r_0}\frac{\partial r_0}{\partial s}. \tag{A.29}$$

A dot above the symbol denotes its material derivative versus time.

The present form of the equations (A.27), (A.29) makes it possible to determine stressed and stained state of the shell with respect to its current position. The equations were derived under assumptions that the shear strains were infinitely small. Nevertheless, displacements and normal components of strains could be final. Displacements vary along the shell thickness in the following way:

$$\tilde{u}(s,z,t) = u(s,t) - z\varphi(s,t); \quad \tilde{w}(s,z,t) = w(s,t).$$

The boundary conditions at the axis of symmetry ($r_0 = 0$) have the following form:

$$u = 0, \quad \varphi = 0, \quad \rho\frac{\partial^2 w}{\partial t^2} = 2\frac{\partial Q}{\partial s} - \left(\frac{T_1}{R_1} + \frac{T_2}{R_2}\right) - \frac{1}{\delta}\left[P_n\left(\delta/2\right) - P_n\left(-\delta/2\right)\right]. \tag{A.30}$$

In deriving (A.30) it was taken into account, that $P_s\left(\pm\delta/2\right) = 0$ and $\dfrac{d\delta}{ds} = 0$ at the axis of symmetry.

The model elastoviscoplastic media (Perzyna, 1968) will be used as constitutive equations, which have the following form in a general three-dimensional case:

$$\dot{e}_{ij} = \frac{\dot{S}_{ij}}{2\mu} + \frac{\dot{S}_{ij}}{2\eta} \cdot \frac{S_n - \sqrt{2/3}J_0}{S_n} H\left(S_n - \sqrt{2/3}J_0\right). \qquad (A.31)$$

The following notations were adopted in (A.31): $S_{ij} = \sigma_{ij} - \dfrac{1}{3}\sigma_{kk}\delta_{ij}$ — components of the deviator of stress tensor; $S_n = \sqrt{S_{ij}S_{ij}}$ — loading intensity; $\dot{e}_{ij} = \dot{\varepsilon}_{ij} - \dfrac{1}{3}\dot{\varepsilon}_{kk}\delta_{ij}$ — components of the deviator of strain rates tensor; $\delta_{ij} = \{0,\ i \neq j;\ 1,\ i = j\}$; μ — shear modulus; η — dynamic viscosity of the material of the shell; J_0 — static limit of plasticity in tension; $H(x)$ — a unit Heavside function ($H(x) = 0,\ x \leq 0$; $H(x) = 1,\ x > 0$); $i, j = 1, 2, 3$.

For spherical parts of stress (σ_{ij}) and strain (ε_{ij}) tensors the Perzyna model adopts laws typical for solid deformable bodies:

$$\sigma = K\varepsilon \qquad (A.32)$$

where K is a volume modulus; $\sigma = \dfrac{1}{3}\sigma_{kk}$, $\varepsilon = \dfrac{1}{3}\varepsilon_{kk}(k = 1,2,3)$.

The axisymmetrical case being regarded as an example the only three components of the strain rate tensor differ from zero ($\dot{\varepsilon}_1, \dot{\varepsilon}_2, \dot{\varepsilon}_{1z}$) and only four components of the stress tensor ($\sigma_{11}, \sigma_{22}, \sigma_{1z}, \sigma_{zz}$). Then the equations (A.31), (A.32) take the form:

$$\dot{S}_{11} + \frac{\mu}{\eta}S_{11} \cdot \frac{S_n - \sqrt{2/3}J_0}{S_n} \cdot H\left(S_n - \sqrt{2/3}J_0\right) = \frac{2}{3}\mu(2\dot{\varepsilon}_1 - \dot{\varepsilon}_2),$$

$$\dot{S}_{22} + \frac{\mu}{\eta}S_{22} \cdot \frac{S_n - \sqrt{2/3}J_0}{S_n} \cdot H\left(S_n - \sqrt{2/3}J_0\right) = \frac{2}{3}\mu(2\dot{\varepsilon}_1 - \dot{\varepsilon}_2)$$

$$\dot{\sigma}_{1z} + \frac{\mu}{\eta}\sigma_{1z} \cdot \frac{S_n - \sqrt{2/3}J_0}{S_n} \cdot H\left(S_n - \sqrt{2/3}J_0\right) = 2\mu\dot{\varepsilon}_{1z}, \qquad (A.33)$$

$$\sigma = K(\varepsilon_1 + \varepsilon_2),\ \ \sigma_{11} = \sigma + S_{11},\ \ \sigma_{22} = \sigma + S_{22},$$

$$S_n = \sqrt{2(S_{11}^2 + S_{11} \cdot S_{22} + \sigma_{1z}^2)}.$$

The criterion for the beginning of macroscopical destructions within the shell is assumed to be a thermodynamic criterion for accumulation of the specific dissipation above the critical value (Kiselev, Yumashev, 1990):

$$D = \int_0^{t_*} \frac{d}{\rho}dt = D_*, \qquad (A.34)$$

where t_* — the time for breakup beginning; D_* — a material constant (critical specific dissipation), normally determined on the basis of experiments on spallation in impact of flat plates (Kiselev, Yumashev, 1990); d — specific dissipation determined by the formula

$$d = S_{ij} \dot{\varepsilon}_{ij}^p,$$

where $\dot{\varepsilon}_{ij}^p$ — components of plastic strain rates tensor.

In our case:

$$d = \frac{S_n}{2\eta} \left(S_n - \sqrt{\tfrac{2}{3} J_0} \right) \cdot H \left(S_n - \sqrt{\tfrac{2}{3} J_0} \right). \tag{A.35}$$

As the shell is assumed to be rather thin, a natural definition for the breakup criterion would be surpassing the critical value D_* by the mean dissipation

$$\bar{D} = \frac{1}{\delta} \int_{-\delta/2}^{\delta/2} \left(\int_0^{t_*} \frac{d}{\rho} dt \right) dz \geq D_*. \tag{A.36}$$

On satisfying the criterion (A.36) at any cross-section $s = s_*$ of the shell a crack should start in the zone causing desintegration of the shell into fragments.

The present mathematical model for dynamical deforming and breakup of a shell could be broadened easily by incorporating in it models for damageable materials, taking into account accumulation of microdamages and defects at a predestruction stage (Kiselev, Yumashev, 1990; Kiselev, 1998).

REFERENCES

Chapter 1

1. Chobotov V.A. (ed.). *Orbital Mechanics (2nd ed.)*, AIAA Education Series. Washington, D.C., 1996.
2. *Technical Report on Space Debris*, (ISBN 92-1-100813-1). United Nations, New York, 1999.
3. *The Orbital Debris Quarterly News*, NASA, 1(2), September 1996.
4. Kessler D.J., Reynolds R.C., Anz-Meador P.D. Orbital Debris Environment for Spacecraft Designed to Operate in Low Earth Orbit. *NASA Technical Memorandum* 100 471, April 1989.
5. Kessler D. Derivation of the Collision Probability Between Orbiting Objects: The Lifetimes of Jupiter's Outer Moons. *Icarus*, **48**, 39–48, 1981.
6. Vedder J.D., Tabor J.L. New Method for Estimating Low Earth Orbit Collision Probabilities. *Journal of Spacecraft and Rockets*, **28**(2), 210–215, March–April 1991.
7. Chobotov V.A., Johnson C.G. Effects of Satellite Bunching on the Probability of Collision Geosynchronous Orbit. *Journal of Spacecraft and Rockets*, **31**(5), 895–899, September–October 1994.
8. Chobotov V.A., Herman D.E., Johnson C.G. Collision and Debris Hazard Assessment for a Low-Earth-Orbit Space Constellation. *Journal of Spacecraft and Rockets*, **34**(2), 233–238, March–April 1997.
9. Chobotov V.A., Mains D.L. Tether Satellite Systems Collision Study. 49th Congress of IAF, Sept. 28–Oct. 2, 1998. Melbourne, Australia (IAA-98-IAA.6.6.02).
10. Jenkin A.B. Probability of Collision During the Early Evolution of Debris Clouds. IAA, 95-IAA.6.4.03. Oslo, Norway.
11. Chobotov V.A. Dynamics of Orbital Debris Clouds and the Resulting Collision Hazard to Spacecraft. *Journal of the British Interplanetary Society*, **43**, 187–195, May 1990.
12. Chobotov V.A. *et al.* Dynamics of Debris Motion and the Collision Hazard to Spacecraft Resulting from an Orbital Breakup. The Aerospace Corp., El Segundo, CA.

13. Chobotov V.A., Spencer D.B. Debris Evolution and Lifetime Following an Orbital Breakup. *Journal of Spacecraft and Rockets*, **28**(6), 670–676, Nov.–Dec. 1991.

14. Jenkin A.B. DEBRIS: A Computer Program for Debris Cloud Modeling. 44th Congress of IAF, Oct. 16–22, 1993, Graz, Austria (AIAA 6.3-93-746).

15. Su S.Y., Kessler D.J. Contribution of Explosion and Future Collision Fragments to the Orbital Debris Environment. COSPAR, Graz, Austria, June 1984.

16. McKnight D.S. Determination of Breakup Initial Conditions. 29th Aerospace Sciences Meeting, AIAA Paper 91-0299, Reno, NV, Jan. 1991.

17. Sorge M.E., Johnson C.G. Space Debris Hazard Software: Program IMPACT Version 3.0 User's Guide. Aerospace Corp. TOR-93(3076)-3, Aug. 1993.

18. Wiedeman D. *et al.* Debris Modeling of Liquid Metal Droplets Released by RORSATS', presented at the 49th IAF Congress in Melbourne, Australia in September, 1998, Paper No. IAA-98-IAa.6.3.03.

19. Spencer D.B. *et al.* Current Space Debris Research in the U.S. Department of Defense, presented at the Seventh International Space Conference of Pacific-Basin Societies, 15–18 July, 1997 in Nagasaki, Japan.

20. AIAA, MEO/LEO Constellations: US Laws, Policies, and Regulations on Orbital Debris Mitigation. SP-016-2-1999, 1999.

21. IAA Position paper on Orbital Debris. International Academy of Astronautics, 1995 (ISBN 2-9508153-1-6).

Chapter 2

1. ESA Space Debris Working Group: Space Debris – The Report of the ESA Space Debris Working Group, ESA SP-1109, Nov. 1988.

2. W. Flury (ed.): *Proceedings of the First European Conference on Space Debris*, ESA SD-01, July 1993.

3. B. Kaldeich, B. Harris (eds.): *Proceedings of the Second European Conference on Space Debris*, ESA SP-393, May 1997.

4. W. Flury (ed.): *Space Debris*, **16**(11), *Advances in Space Research*. Pergamon, 1995.

5. W. Flury (ed.): *Space Debris*, **19**(2), *Advances in Space Research*. Pergamon, 1997.

6. W. Flury, H. Klinkrad (eds.): *Space Debris*, **23**(1), *Advances in Space Research*. Pergamon, 1999.

7. R. Jehn, S. Vinals-Larruga, H. Klinkrad. DISCOS – The European Space Debris Database, 44th IAF Congress, 16–22 Oct. 1993. Graz, Austria.

8. D.G. King-Hele, D.M.C. Walker, A.N. Winterbottom, J.A. Pilkington, H. Hiller, G.E. Perry. The RAE Table of Earth Satellites 1957–1989, RAE, Farnborough, England, 1990.

9. H. Klinkrad, W. Flury, R. Jehn, G. Drolshagen, R. Czichy. European Efforts in Modeling the Space Debris Environment, 19th ISTS, Yokohama, 1994.

10. W. Flury. The Space Debris Environment of the Earth. *Earth, Moon and Planets*, **70**, 79–91, 1995.

11. H. Klinkrad. Description and Forecast of the Terrestrial Space Debris Environment, ESA Conference on Spacecraft Structures, Materials and Mechanical Testing, ESTEC, Noordwijk, March 1996.

12. H. Klinkrad, J. Bendisch, H. Sdunnus, P. Wegener, R. Westerkamp. An Introduction to the 1997 ESA MASTER Model, Second European Conference on Space Debris, ESA SP-393, 1997.

13. P. Wegener *et al.* The Orbital Distribution and Dynamics of Solid Rocket Motor Particle Clouds for an Implementation into the MASTER Debris Model, *Adv. of Space Res.*, **23**(1), 161–164, 1999.

14. E. Gruen, H.A. Zook, H. Fechtig, R.H. Giese. Collisional Balance of the Meteoritic Complex, *Icarus*, **62**, 244–272, 1985.

15. P. Staubach, E. Gruen. Upgrade of the DISCOS meteoroid model, ESOC Contract No. 10463/93, Max-Planck-Institute for Nuclear Physics, Heidelberg, 1996.

16. L. Anselmo, A. Cordelli, P. Farinella, C. Pardini, A Rossi. Long-Term Evolution of Earth Orbiting Debris, ESOC Contract No. 10034/92, Consorzio Pisa Ricerche, Pisa, Italy, 1996.

17. J. Bendisch, D. Rex. Analysis of Debris Mitigation Measures, ESOC Contract No. 11263/95, Technical University of Braunschweig, Braunchweig, Germany, 1996.

18. G. Drolshagen, J.A.M. McDonnell, T.J. Stevenson, S. Deshpande, L. Kay, W.G. Tanner, J.C. Mandeville, W.C. Carey, C.R. Maag, A.D. Griffiths, N.G. Shrine, R. Aceti. Optical survey of micrometeoroid and space debris impact features on EURECA, *Planet and Space Sci.*, **44**(4), 317–340, 1996.

19. J.A.M. McDonnell, A.D. Griffiths, G. Drolshagen, J.C. Mandeville, W.C. Carey. Meteoroid and debris impact environment of the Hubble Space Telescope solar arrays – first results. *Proc. of the Hubble Space Telescope Solar Array Workshop*, ESTEC, WPP-77, pp. 501–509, 1995.

20. C.R. Maag *et al.* The contamination environment at the MIR Space Station as measured during the EUROMIR '95 mission, *Proc. 7-th Int. Symposium "Materials in Space Environment"*, ESA SP-399, pp. 301–308, 1997.

21. G. Drolshagen *et al.* Microparticles in the geostationary orbit (GORID experiment), *Adv. of Space Res.*, **23**(1), 123–133, 1999.

22. D. Mehrholz. Radar Detection of Mid-Size Debris, *Adv. of Space Res.*, **16**(11), 17–27, 1995.

23. D. Mehrholz, L. Leushacke, R. Jehn. The COBEAM-1/96 Experiment, *Adv. of Space Res.*, **23**(1), 23–32, 1999.

24. L. Leushacke, D. Mehrholz, R. Jehn. First FGAN/MPIfR cooperative debris observation campaign: experiment outline and first results, Second European Conference on Space Debris, ESA SP-393, 1997.

25. M. Lambert. Shielding against natural and man-made debris, a growing challenge. *ESA. J.*, **17**(1), 31–42, 1993.

26. A. Stilp. Hypervelocity impact research, Second European Conference on Space Debris, ESA SP-393, 1997.

27. M. Sanchez, C. Bonnal, W. Naumann. Ariane debris mitigation measures, Second European Conference on Space Debris, ESA SP-393, 1997.

28. G. Janin. Log of Objects near the Geostationary Ring, Issue **19**, ESOC, Feb. 1999.
29. W. Flury, H. Heusmann, W. Naumann. Cost Effectiveness of Debris Mitigation Measures, 46th IAF Congress, Oslo, 1995.
30. H. Klinkrad. The ESA Space Debris Mitigation Handbook, Rel. 1.0, ESOC, Darmstadt, Feb. 1999.

Chapter 3

1. Chobotov V.A. Dynamics of orbital debris clouds and the resulting collision hazard to spacecrafts. *FBIS, Journal of the British Interplanetary Society*, **43**(5), 187–195, 1990.
2. Flury W. Space debris: a European view. *Proceedings of Space Debris Forum, JSASS*, pp. 37–54. Tokyo, May 15, 1992.
3. Kessler D.J. Collisional cascading: the limits of population growth in low Earth orbit. *Advances in space Res*, **11**(12), 1991.
4. Loftus J.P., Anz-Meador P.D., Reynolds R. Space debris minimization and mitigation plans and practices. *Space Debris Forum, JSASS*, pp. 16–27. Tokyo, May 15, 1992.
5. Yasaka T., Ishii N. Breakup in geostationary orbit: a possible creation of a debris ring. *IAA-91-596. 42nd Congress of the International Astronautical Federation*, pp. 1–12. Montreal, October 5–11, 1991.
6. Perek L. Space debris as an issue for the international community. *Space Debris Forum, JSASS*, pp. 1–15. Tokyo, May 15, 1992.
7. N. Smirnov, V. Dushin, I. Panfilov, V. Lebedev. Space debris evolution mathematical modeling. *Proc. 1st European Conf. on Space Debris*, ESA-SD-01, pp. 309–316. Darmstadt, 1993.
8. Reynolds R. *Documentation of Program EVOLVE:* A Numerical Model to Compute Projections of the Man-Made Orbital Debris Environment. *System Planning Corp. Tech. Rpt.* OD91-002-U-CSP, 1991.
9. *Technical Report on Space Debris*, United Nations Organization, A/AC, 105/720. New York, 1999.
10. Ivanov V.L., Menshikov V.A., Pchelintsev L.A., Lebedev V.V. *Space Debris*, p. 360. Moscow, Patriot Pbl., 1996.
11. Tshernyavsky G.M., Nazarenko A.I. Simulation of the near-Earth Space Contamination. *"Collisions in the Surrounding Space"*, edited by A.G. Masevich, pp. 104–129. Moscow, Kosmoinform, 1995.
12. Portee D.S.F., Loftus J.P. *Orbital Debris: a Chronology*. NASA/TP-1999-208856. Houston, 1999.
13. Krisko P. Evolve 4.0 Preliminary Results. *Orbital Debris Quarterly News*, **4**(1), 6–7, 1999.
14. Bendisch J., Klinkrad H., Krag H., Rex D., Sdunnus H., Wegener P., Wiedemann C. Results of the Upgraded Master Model. *50-th IAF Congress*, Amsterdam, 1999, IAA-99-IAA.6.4.05.

15. Walker R., Stokes P.H., Wilkinson J.E., Swinerd G.G. Long-Term Collision Risk Prediction for Low Earth Orbit Satellite Constellation. *50-th IAF Congress*, Amsterdam, 1999, IAA-99-IAA.6.6.04.

16. Kasimenko T.V., Ryhlova L.V. the upper atmosphere as a Space Debris Cleaner. *"Collisions in the Surrounding Space"*, edited by A.G. Masevich, pp. 169–172. Moscow, Kosmoinform, 1995.

17. Poljakhova E.N. The Role of Light Pressure in Astronomy and Space Research. *"Collisions in the Surrounding Space"*, edited by A.G. Masevich, pp. 173–251. Moscow, Kosmoinform, 1995.

18. Mikisha A.M., Smirnov M.A. Secular Evolution of Space Bodies on High Orbits due to Light Pressure. *"Collisions in the Surrounding Space"*, edited by A.G. Masevich, pp. 252–271. Moscow, Kosmoinform, 1995.

19. Khutorovsky Z.N., Kamensky S.Yu., Boikov V.E., Smelov V.L. Collisions of Space Objects in Low Orbits. *"Collisions in the Surrounding Space"*, edited by A.G. Masevich, pp. 19–90. Moscow, Kosmoinform, 1995.

20. Rossi A., Auselmo L., Pardini C., Cordelli A., Farinella P., Parinello T. Approaching the Experimental Growth: Parameter Sensitivity of the Debris Evolution. *Proc. First Europ. Conf. on Space Debris*, ESASD-01, pp. 287–292. Darmstadt, 1993.

21. Rex D., Eichler P. The Possible Long Term Overcrowding of LEO and the Necessity and Effectiveness of Debris Mitigation Measures. *Proc. First Europ. Conf. on Space Debris*, ESASD-01, pp. 607–615. Darmstadt, 1993.

22. Anz-Meador P.D., Potter A.E. Density and Mass Distribution of Orbital Debris. *Acba Astronaution*, **38**(12), 927–936, 1996.

23. Matney M.J., Theall J.R. The Use of the Satellite Breakup Risk Assessment Model (SBRAM) to Characterize Collision Risk to Manmed Spacecraft. *50-th IAF Congress*, IAA-99-IAA.6.5.09. Amsterdam, 1999.

24. N. Smirnov, V. Nikitine, A. Kiselev. Peculiarities of Space Debris Production in Different Types of Orbital Break-ups. *Proc. 2nd European Conf. on Space Debris*, pp. 465–471. Darmstadt, 1997.

25. N. Smirnov, A. Kiselev, V. Lebedev. Mathematical Modeling of Space Debris Evolution in Low Earth Orbits. *Proc. 19th Int. Symposium on Space Technology and Science*, 94-n-28p. Yokohama, 1994.

26. A. Kiselev. Simple Mathematical Models for Spacecraft Fragmentation in Explosion. *Journal of Applied Mech. And Tech. Physics*, **2**, 159–165, 1995.

27. N. Smirnov, V. Nikitine, A. Kiselev. Gas Explosions in Confined Volumes. *Proc. Int. Symposium on Hazards, Prevention and Mitigation of Industrial Explosions*. Christian Michelsen Research AS, Bergen, **2**, 515–561, 1996.

28. Chobotov V.A., Spencer D.B. Debris Evolution and Lifetime Following an Orbital Breakup. *Journ. Spacecraft*, **28**(6), 670–676, 1991.

29. Talent D. Analytic Model for Orbital Debris Environmental Management. *AIAA Paper No 1363*, 1–10, 1990.

30. Nazirov R.R., Ryasanova E.E., Sagdeev R.Z., Sukhanov A.A. Analysis of Space Self-Cleaning of Debris due to Atmospheric drag. *Space Research Institute Pr-1670*. Moscow, 1990.

31. Nazarenko A. The Development of the Statistical Theory of a Satellite Ensemble Motion and its Application to Space Debris Modeling. *Proc. 2nd European Conf. on Space Debris*. Darmstadt, 1997.

32. Lim I. Orbital Debris Evolution Modeling Accounting for Self-Production. Diploma paper. Gas and Wave Dynamics Dept. of Moscow State Univ., 1995.

33. Interagency Report on Space Debris, OSTP, 1995.

Chapter 4

1. Draft Report of the Scientific and Technical Subcommittee on the Work of its Thirty-sixth Session (UNCOPUOS). A/AC.105/707, December 98.

2. R. Reynolds. Documentation of Program EVOLVE: A Numerical Model to Compute Projections of the Man-Made Orbital Debris Environment, System Planning Corp. Technical Report OD91-002-U-CSP, February, 1991.

3. D. Kessler. Derivation of the collision probability between orbiting objects: The lifetime of Jupiter's outer moons. *Icarus*, **48**, 1981.

4. A. Bade. Orbital Debris Quarterly News. NASA, Johnson Space Center, September 1996, **1**(2).

5. D. Talent. Analytic model for orbital debris environmental management. AIAA Paper, No 1363, 1990.

6. V.L. Ivanov, V.A. Menshikov, L.A. Pchelintsev, V.V. Lebedev. *Space Debris. A Problem and Ways of its Solving*, v.1. Moscow, Patriot, 1996.

7. R. Jehn, A. Nazarenko *et al*. Comparison of Space Debris in the Centimeter Size Range. Second European Conference on Space Debris, ESOC, Darmstadt, Germany, March 1997.

8. A. Nazarenko. The Development of the Statistical Theory of a Satellite Ensemble Motion and its Application to Space Debris Modeling. Second European Conference on Space Debris, ESOC. Darmstadt, Germany, 17–19 March 1997.

9. A. Nazarenko. A Model of Distribution Changes of the Space Debris. *The Technogeneous Space Debris Problem*. Moscow, COSMOSINFORM, 1993.

10. A. Nazarenko, Comparison of the Space Debris (SD) Modeling Data with Haystack Radar Measurements. 12th IADC Meeting, NASA/JSC. Houston, 8–10 March 1995.

11. E. Stansbery, M. Matney. Recent Haystack Results. 13th Inter-Agency Space Debris Coordination Meeting, ESA/ESOC. Darmstadt, 28 Feb–01 Mar, 1996.

12. L. Leushacke. Mid-Size Space Debris Measurements with the TIRA System. 48th International Astronautical Congress, IAA-97-IAA.6.3.02. October 1997.

13. A. Nazarenko. Contribution of Elliptical Orbits into Space Debris Environment. 15th IADC Meeting. Houston, December 9–12, 1997.

14. A. Nazarenko. The Altitude-Latitude Space Debris Distribution. *The Technogeneous Space Debris Problem*. Moscow, COSMOSINFORM, 1993.

15. D. Kessler and B. Cour-Palais. Collision Frequency of Artificial Satellites: *The Creation of a Debris Belt. Journal of Geophysical Research*, **83**(A6), 2637–2645, 1978.

16. Z. Khutorovsky, S. Kamensky, Direct method for the analysis of collision probability of artificial space objects in LEO: techniques, results and applications. First European Conference on Space Debris. Darmstadt, Germany, April 1993.

17. D. Kessler. Collision Probability at Low Altitudes Resulting from Elliptical Orbits. *Adv. Space Res.*, **10**(3–4), 1990.

18. D. Kessler. Collision Cascading: The Limits of Population Growth in Low Earth Orbit. *Adv. in Space Res.*, **11**(12), 1991.

19. A. Rossi, P. Farinella, Collision Rates and Impact Velocities for Bodies in Low Earth Orbit. *esa Journal 92/3.*

20. Z. Khutorovsky *et al.* Collision of Space Objects on Low Orbits. *Collisions in the Surrounding Space (Space Debris).* Moscow, COSMOSINFORM, 1995.

21. A. Nazarenko. Aerodynamic Analogy for Interactions between Spacecraft of Different Shapes and Space Debris. *Cosmic Research*, **34**(3), 1996.

22. E.L. Christiansen. Working Group#3 – Protection. 16th IADC Meeting. Toulouse, France, November 1998.

23. D. Kessler *et al.* A Computer-Based Orbital Debris Environment Model for Spacecraft Design and Observation in Low Orbit, NASA, Technical Memorandum 104825, November 1996.

Chapter 5

1. Flight International, pp. 338–339, 27 Aug. 1964.

2. CCIR Document 4/141-E, 17 June, 1992.

3. L. Perek. Safety in the Geostationary Orbit After 1988. IAA-89-632, 40th Congress of IAF. Malaga, 1989.

4. V.A. Chobotov, M.G. Wolfe. End-of-Life Disposal of Spacecraft in Geosynchronous Orbit. IAA-89-631, 40th Congress of IAF, Malaga, 1989.

5. T. Yasaka, S. Oda. Classification of Debris Orbits with Regard to Collision Hazard in Geostationary Region. *Space Safety and Rescue 1990*, edited by G. Heath, pp. 170–184, AAS Publication (1991).

6. T. Yasaka, N. Ishii. Breakup in Geostationary Orbit: A Possible Creation of a Debris Ring. *Space Safety and Rescue 1991*, edited by G. Heath, pp. 203–216, AAS Publication (1993).

7. T. Yasaka, T. Hanada, H. Hirayama. "GEO Debris Environment: A Model to Forecast the Next 100 Years", *Adv. In Space Res.*, **23**(1), 191–199, 1999.

8. S.Y. Su, D.J. Kessler. Contribution of Explosion and Future Collision Fragments to the Orbital Debris Environment. *Adv. In Space Res.*, **5**(2), 234, 1985.

9. T. Yasaka, T. Hanada. Low Velocity Impact Test and Its Implications to Object Accumulation Model in GEO. *Adv. Astronautical Sciences*, **91**, 1029–1038, 1996.

10. T. Hanada, T. Yasaka, K. Goto. Fragments Creation via Impact at Low Speed. *Adv. Astronautical Sciences*, **96**, 979–986, 1997.

11. H. Klinkrad, J. Bendisch, H. Sdunnus, P. Wegener, R. Westerkamp. An Introduction to the 1997 ESA MASTER Model, Proc. 2nd European Conference on Space Debris, pp. 217–224. Darmstadt, 1997 (ESA SP-393).

12. N. Johnson, E. Christiansen, R. Reynolds, M. Matney, J.C. Zhang, P. Eichler, A. Jackson. NASA/JSC Orbital Debris Models, ibid, pp. 222-232.

13. P. Eichler, D. Rex. Chain Reaction of Debris Generation by Collisions in Space – A Final Threat to Spaceflight? IAA-89-628, 40th IAF. Malaga, 1989.

14. T. Yasaka, S. Oda. Breakup, Collision and Population Growth in Geostationary Orbit. Proc. 1st European Conference on Space Debris, pp. 646–649. Darmstadt, 1993 (ESA SD-01).

15. T. Yasaka. Debris Population Growth in Geostationary Orbit. Proc. 19th ISTS, pp. 979–984. Yokohama, 1994.

16. T. Yasaka. Large Orbital Object Removal – Its Necessity and Technological Options — Proc. 1st European Conference on Space Debris, pp. 596-600. Darmstadt, 1993 (ESA SD-01).

Chapter 6

1. Bess, Dale T. Mass distribution of orbiting man-made space debris. NASA Technical Note D8108, December 1975.

2. Reynolds R.C. Documentation of Program EVOLVE: A Numerical Model to Compute Projections of the Man-Made Orbital Debris Environment. SPC Rep, No. OD91-002-U-CSP, System Planning Corp., Nassau Bay, Texas, USA (I 5 February 1991); and Reynolds R.C. Review of Current Activities to Model and Measure the Orbital Debris Environment in Low-Earth Orbit. *Adv. Space Res.*, **10**(3–4), (3)359–(3)371, 1990.

3. Eichler P., Rex D. Debris chain reactions. *Orbital Debris: Technical Issues and Future Directions*, NASA Conference Publication 10077, pp. 187–195, 1992.

4. Badhwar G., Anz-Meador P.D. Relationship of radar cross sections to the geometric size of orbital debris. *Orbital Debris: Technical Issues and Future Directions*, NASA Conference Publication 10077, pp. 150–153. 1992.

5. Badhwar G.D., Anz-Meador P.D. Mass estimation in the breakups of Soviet satellites Journal of the British Interplanetary Society, **40**, 403–410 (1990).

6. Nauer, D.J. The Fragmentation of Cosmos 1484. Colorado Springs, CO: Teledyne Brown Engineering report CS94-LKD-003, 17 November 1993.

7. Anz-Meador P.D., Rast R.H. and Potter A.E. Apparent Densities of Orbital Debris. *Adv. Space Res.* **13**(8), (8)153–(8)156, 1993.

8. Anz-Meador P.D., Henize K.G. and Kessler D.J. The NASA Catalog of Adjusted U.S. Space Command RCS'S, NASA JSC-26738, NASA L. B. Johnson Space Center, Houston, Texas, USA, in press (July 1994).

9. David A.H., Hartley H.O. and Pearson E.S. The distribution of the ratios, in a single normal sample, of range to standard deviation. *Biometrika* **34** (1954): 482–493. Referenced in Applied Statistics: A Handbook of Techniques, edited by Lother Sachs. New York: Springer Verlag, 1984.

10. Bohannon G.D. Comparisons of Orbital Debris Size Estimation Methods Based on Radar Data. Xontech Rep, 920123-BE-2048, Xontech, Inc. Los Angeles, California, USA (May 1992).

11. Grissom I.W., Guy R.P. and Nauer D.J. *History of On-Orbit Satellite Fragmentations*, 8th ed., Colorado Springs, CO: Teledyne Brown Engineering Report CS94-LKD-018, July 1994.

12. Brent R.P. *Algorithms for Minimization without Derivatives*, p. 78, Prentice-Hall (1973). In *Numerical Recipes; the Art of Scientific Computing*, William H. Press *et al.*, p. 283 ff. Cambridge University Press (1987).

13. Bohannon G., Dalquist C. and Pardee M. Physical Description of Debris Objects Used in Static RCS Measurements. Xontech Rep. 910555-1978, Xontech, Inc., Los Angeles, California, USA (August 1991).

14. Hogg D.M., Cunningham T.M. and Isbell W.M. Final Report on the SOCIT Series of Hypervelocity Impact Tests. Rep. WL-TR-93-7025, Wright Laboratory Armament Directorate, US Air Force Eglin Air Force Base (July 1993).

15. Fucke W. and Sdunnus H. Population Model of Small Size Space Debris. BF-R 67.698-5, Battelle-Institut e.V., Frankfurt, Germany (June 1993).

16. Sato T., Wakayama T., Tanaka T., Ikeda K.-I. and Kimura I. Shape of Space Debris as Estimated from Radar Cross Section Variations. *J. Spacecraft Rockets*, **31**, 665–670 (1994).

Chapter 7

1. Chobotov V.A. Dynamics of orbital debris clouds and the resulting collision hazard to spacecrafts. FBIS, *Journal of the British Interplanetary Society*, **43**(5), 187–195, 1990.

2. Flury W. Space debris: a European view. Proceedings of Space Debris Forum, JSASS, pp. 37–54. Tokyo, May 15, 1992.

3. Kessler D.J. Collisional cascading: the limits of population growth in low Earth orbit. *Advances in space Res.*, **11**(12), 1991.

4. Loftus J.P., Anz-Meador P.D., Reynolds R. Space debris minimization and mitigation plans and practices. Space Debris Forum, JSASS, pp. 16–27. Tokyo, May 15, 1992.

5. Yasaka T., Ishii N., Breakup in geostationary orbit: a possible creation of a debris ring. IAA-91-596. 42nd Congress of the International Astronautical Federation, pp. 1–12. Montreal, October 5–11, 1991.

6. Christiansen E.L., Horn J.R., Crews J.L. Augmentation of Orbital Debris Shielding for Space Station Freedom. AIAA 90-3665, Huntsville, AL, 1990.

7. Talent D. Analytic model for orbital debris environmental management. AIAA Paper No. 1363, pp. 1–10, 1990.

8. Christiansen E.L. Performance Equations for Advanced Orbital Debris Shields. AIAA 92-1462, Huntsville, AL, 1992.

9. Nazarenko A.I., Prediction and Analysis of Orbital Debris Environment Evolution. Proc. of the First European Conference on Space Debris ESA-SD-01, pp. 293–298. Darmstadt, 1993.

10. Smirnov N.N., Dushin V.R., Panfilov I.I., Lebedev V.V. Space Debris Evolution Mathematical Modeling. Proc. of the First European Conference on Space Debris ESA-SD-01, pp. 309–316. Darmstadt, 1993.

11. Chobotov V.A., Spencer D.B. Debris Evolution and Lifetime Following an Orbital Breakup. *Journal Spacecraft*, **28**(6), 670–676, 1991.

12. McKnight D.S., Nagl L. Key Aspects of Satellite Breakup Modeling. Proc. of the First European Conference on Space Debris ESA-SD-01, pp. 269–274. Darmstadt, 1993.

13. Loftus J.R. (ed.) Orbital Debris from Upper Stage Breakup. AIAA Proceedings of the Conference, 1987.

14. Inventory of Orbiting Hypergolic Rocket Stages. Technical memorandum SN-3-81-55, Technical planning office, NASA-JSC, March, 1981.

15. Webster I.J., Kawamura T.Y. Precluding Post-Launch Fragmentation of Delta Stages. McDonnel Douglas Space Systems Company Tech. Rept., 5.05.92.

16. Smirnov N.N., Nikitin V.F., Kiselev A.B. Gas Explosions in Confined Volumes. Proc. International Symposium on Hazards, Prevention and Mitigation of Industrial Explosions. Christian Michelsen Research AS, **2**, 515–561. Bergen, Norway, 1996.

17. Smirnov N.N., Lebedev V.V., Kiselev A.B. Mathematical Modeling of Space Debris Evolution in Low Earth Orbits. 19-th ISTS, 94-N-28 p. Yokohama, Japan, 1994.

18. Fucke W. Fragmentation Experiments for the Evaluation of the Small Size Debris Population. Proc. of the First European Conference on Space Debris ESA-SD-01, pp. 275–280. Darmstadt, 1993.

19. Investigation of Delta Second Stage On-Orbit Explosions. Report MDC-H0047, McDonnel Douglas Astronautics Company Huntington Beach, CA, 1982.

20. Smirnov N.N., Panfilov I.I. Deflagration to Detonation Transition in Combustible Gas Mixtures. *Combustion and Flame*, **101**(1/2), 91–100. Elsevier Publ. Co., 1995.

21. Smirnov N.N., Tyurnikov M.V. A Study of Deflagration and Detonation in Multiphase Hydrocarbon-Air Mixtures. *Combustion and Flame*, **96**, 130–140. Elsevier Publ. Co., 1994.

22. Smirnov N.N. *et al.* Theoretical and Experimental Investigation of Combustion to Detonation Transition in Chemically Active Gas Mixtures in Closed Vessels. *Journal of Hazardous Materials*, **53**, 195–211, 1997.

23. Kiselev A.B., Yumashev M.V. Deformation and Fracture under Impact Loading Model of Damaged Thermoelastoplastic Medium. *Appl. Mech. Tech. Phys.*, **31**(5), 116–123, 1990.

24. Kiselev A.B. The Model of Thermoelastoplastic Deformation and Fracture of Materials under Multiaxial Loading. Fourth Int. Conf. On Biaxial/Multiaxial Fatigue, St. Germain en Laye, France, 1994.

25. Kiselev A.B. Mathematical Modeling of Dynamical Deforming and Combined Microfracture of Damageable Thermoelastoviscoplastic Medium. *Advances Methods in Material Processing Defects (SAM Series)*, edited by M. Predeleanu and P. Gilormini. Elsevier Science B.V., Amsterdam, The Netherlands, 1997.

26. Su S.-Y., Kessler D.J., Contributions of Explosions and Future Collision Fragments to the Orbital Debris Environment. *Advances in Space Research*, **5**(2), 25–34, 1985.

27. Potter A., Anz-Meador P.D. *et al*. Rept. At 45-th Congress of the IAF, IAA-94, Jerusalem, Israel, Oct. 9–14, 1994.

28. Smirnov N.N., Zverev N.I., Tyurnikov M.V. Two-phase Flow Behind a Shock Wave with Phase Transitions and Chemical Reactions. *Experimental Thermal and Fluid Science*, **13**(1), 1996.

29. Smirnov N.N., Zverev I.N. *Heterogeneous Combustion*, 446 p. ISBNS-211-01563-0. Moscow University Publ. Co., Moscow, 1992.

30. Kiselev A.B. Mathematical Modeling of Fragmentation of Thin Walled Spherical Sheets under the Action of Dynamical Internal Pressure. *Moscow Univ. Mechanics Bulletin*, **51**(3), 52–60, 1996.

31. Interagency Rpt. On Orbital Debris, OSTP, 1995.

32. Smirnov N.N., Nikitin V.F., Legros J.C. Turbulent Combustion of Multiphase Gas-Particles Mixtures. Thermogravitational Instability. *Advanced Computation & Analysis of Combustion*, edited by G. Roy, S.M. Frolov, P. Givi), pp. 136–160. ENAS Publ., Moscow, 1997.

33. Smirnov N.N., Nikitin V.F. Unsteady-State Turbulent Diffusive Combustion in Confined Volumes. *Combustion and Flame*, **111**, 222–256, 1997.

34. Smirnov N.N., Kiselev A.B., Nikitin V.F. Peculiarities of Space Debris Production in Different Types of Orbital Breakups. Proc. 2nd Europ. Conf. on Space Debris, SP-393, pp. 465–472. Darmstadt, 1997.

35. Anz-Meador P.D., Potter A.E. Density and Mass Distribution of Orbital Debris. *Acta Astronautica*, **38**(12), 927–936, 1996.

36. Smirnov N.N., Tyurnikov M.V. Experimental investigation of Deflagration to Detonation Transition in Hydrocarbon-Air Mixtures. *Combustion and Flame*, **100**, 661–668, 1995.

37. Ivashnyov O.E., Ivashreva M.N., Smirnov N.N. Slow Waves of Boiling under Hot Water Depressurization. *J. Fluid Mech.*, **413**, 149–180, 2000.

38. Cour-Palais B.G., Crews J.L. A Multi-shock Concept for Spacecraft Shielding. *Int. J. Impact. Engng.*. Vol. 10, pp. 135–146, 1990.

Annex

1. Launder B.E. and Spalding D.B. *Mathematical Models of Turbulence*. Academic Press, New York, 1972.

2. Smirnov N.N., Nikitin V.F., Legros J.C. Turbulent combustion of multiphase gas-particles mixtures. Advanced Computation and Analysis of Combustion, pp. 136–160. ONR-RFBR, ENAS Publishers, Moscow, 1997.

3. Pironneau O. and Mohammadi B. *Analysis of the K-Epsilon turbulence model*. Mason Editeur, Paris, 1994.

4. Smirnov N.N., Zverev I.N. *Heterogeneous combustion*, 446 p. Moscow University Publishers, Moscow, 1992.

5. Oran E.S. and Boris J.P. *Numerical simulation of reactive flows*. Elsevier, New York, 1987.

6. Smirnov N.N., Nikitin V.F. Unsteady-state turbulent diffusive combustion in the confined volumes. *Combustion and Flame*, **111**, 222–256, 1997.

7. Smirnov N.N. Combustion and detonation in multi-phase media; initiation of detonations in dispersed-film systems behind a shock wave. *Int. Journ. of Heat and Mass Transfer*, **31**(4), 779–793, 1988.

8. Smirnov N.N., Tyurnikov M.V. A study of deflagration and detonation in multi-phase hydrocarbon-air mixtures. *Combustion and Flame*, **96**, 130–140, 1994.

9. Rose M., Roth P., Frolov S.M, Neuhaus M.G., Klemens R. Lagrangian approach for modeling two-phase turbulent reactive flows. Advanced Computation and Analysis of Combustion, pp. 175–194. ONR-RFBR, ENAS Publishes, Moscow, 1997.

10. Dushin V.R., Nikitin V.F., Smirnov N.N. et al. Mathematical modeling of particles cloud evolution in the atmosphare after a huge explosion. Proceedings of the 5th International Colloquium on Dust Explosions, pp. 287–292. Pultusk – Warsaw, 1993.

11. Nigmatulin R.I. *Dynamics of Multiphase Media*, Science Publishers. Moscow, 1987.

12. Azzopardi B.J., Hewitt G.F. Maximum Drop Sizes in Gas-Liquid Flows. *Multiphase Sci. & Tech.*, **9**, 109–204, 1997.

13. Kiselev A.B. Behavior of an Elastic-Plastic Shell of Revolution under Axisymmetric Dynamic Loading. *Moscow Univ. Mech. Bulletin*, **36**(2), 67–71, 1981.

14. Kiselev A.B., Maksimov V.F. Numerical Simulation of Complex Interaction Between an Elastic-Plastic Shell of Revolution and Elastic Filler. *Moscow Univ. Mech. Bulletin*, **37**(1), 63–68, 1982.

15. Kiselev A.B., Maksimov V.F. Numerical Simulation of Complex Interaction Between a Shell of Revolution and Its Filler, with Hlowance for Friction. *Moscow Univ. Mech. Bulletin*, **39**(2), 29–34, 1984.

16. Kiselev A.B., Maksimov V.F. Numerical Modelling of Normal Punching of a Thin Obstacle by a Deformable Body of Revolution. *Mechanics of Solids*, **30**(5), 146–156, 1995.

17. Perzyna P. *Fundamental Problems of Viscoplasticity*. Moscow, 1968 (in Russian).

18. Kiselev A.B., Yumashev M.V. Deforming and Fracture under Impact Loading. The Model of Damageable Thermoelastoplastic Medium. *J. Appl. Mech. Tech. Phys.*, **31**(5), 116–123, 1990.

19. Kiselev A.B. Mathematical Modeling of Dynamical Deformation and Combined Microfracture of a Thermoelastoplastic Medium. *Moscow Univ. Mech. Bulletin.*, **53**(6), 41–48, 1998.

20. Kiselev A.B., Yumashev M.V., Volod'ko O.V. Deforming and Fracture of Metals. The Model of Damageable Thermoelastoplastic Medium. *J. Of Materials Processing Technology.*, **80–81**, 585–590, 1998.

21. Smirnov N.N., Kulchitsky A.V., Non-equilibrium Effects in Liquid Fuel Droplets Combustion. 51st IAF Congress, Rio de Janeiro, 2000, IAF-00-Y.4.09.

22. Philip M., Dissertation "Experimentelle und theoretische Untersuchungen zum Stabilitatsverhalten von Drallflammen mit zentraler Ruckstromzone". Karlsruhe University, 1991.

23. Smirnov N.N., Nikitin V.F., Legros J.C., Ignition and Combustion of Turbulized Dust-Air Mixtures. *Combustion and Flame*, **123**(1/2), 46–67, 2000.

SUBJECT INDEX

Milton Keynes UK
Ingram Content Group UK Ltd.
UKHW051951071024
449327UK00026B/2259